FASTER, BETTER, CHEAPER

NEW SERIES IN NASA HISTORY
Roger D. Launius, Series Editor

Before Lift-off: The Making of a Space Shuttle Crew,
by Henry S. F. Cooper, Jr.

The Space Station Decision: Incremental Politics
and Technological Choice, *by Howard E. McCurdy*

Exploring the Sun: Solar Science since Galileo,
by Karl Hufbauer

Inside NASA: High Technology and Organizational Change
in the U.S. Space Program, *by Howard E. McCurdy*

Powering Apollo: James E. Webb of NASA,
by W. Henry Lambright

NASA and the Space Industry, *by Joan Lisa Bromberg*

Taking Science to the Moon: Lunar Experiments
and the Apollo Program, *by Donald A. Beattie*

Faster, Better, Cheaper: Low-Cost Innovation
in the U.S. Space Program, *by Howard E. McCurdy*

HOWARD E. MCCURDY

FASTER, BETTER, CHEAPER
Low-Cost Innovation in the U.S. Space Program

The Johns Hopkins University Press
Baltimore and London

The Johns Hopkins University Press
2715 North Charles Street
Baltimore, Maryland 21218–4363
www.press.jhu.edu

Library of Congress Cataloging-in-Publication Data

McCurdy, Howard E.
Faster, better, cheaper : low-cost innovation in the U.S. space program /
Howard E. McCurdy.
p. cm. — (New series in NASA history)
Includes bibliographical references and index.
ISBN 0-8018-6720-7 (hardcover)
1. Astronautics—United States—Cost control. 2. Outer space—
Exploration—Cost control. 3. United States. National Aeronautics and
Space Administration. 4. Organizational effectiveness. 5. Astronautics—
Technological innovations. I. Title. II. Series.
TL796.5.U5 M37 2001
629.4′068′1—dc21   00-012361

A catalog record for this book is available from the British Library.

To Paul P. Van Riper

CONTENTS

BOXES

ACKNOWLEDGMENTS

Before humans flew in space, I marveled at paintings of rocket ships and space stations as artists imagined they would look. Growing up, I studied books on astronomy, read my share of science fiction, built small, homemade rockets, and rode the "Rocket to the Moon" at Disneyland. In college, I majored in chemistry until my fascination with government caused me to switch to political science and eventually public administration. As a doctoral student at what is now the Johnson School of Management at Cornell University, my major professor, Paul P. Van Riper, asked me to assist him with a research project on decision making at the National Aeronautics and Space Administration (NASA). The assignment provided me with my first opportunity to pursue a youthful hobby in conjunction with the academic career I had chosen.

More than any other I have done, this book has allowed me to combine my hobby and my vocation. The book examines rather tedious matters of public management and financial administration in the more exciting realm of space exploration. It recounts the development of a special reform, NASA's "faster, better, cheaper" initiative, whose proponents—beginning in 1992—sought to transform the way space flight missions are run.

I am deeply grateful to the people and institutions that allowed me to pursue this subject, permitting me to convert what is essentially a hobby into a scholarly pursuit. The NASA History Office provided financial support for an account of the initiative. I am especially indebted to Roger D. Launius, the NASA Historian, Lori B. Garver, NASA's Associate Administrator for Policy and Plans, and Alan Ladwig, her predecessor. Without their support and encouragement, I would not have started this work. Sylvia K. Kraemer, the previous NASA Historian, encouraged me to start writing books on space policy fifteen years ago and, from her position in the Office of Policy and Plans, supported this work as well. In the History Office, Nadine Andreassen, Jane Odom, Stephen J. Garber, Colin Fries, and Mark Kahn treated my requests with patience and kindness. Richard Faust assisted from the NASA headquarters library and Johanna A. Gunderson from the NASA comptroller's office, helping me to fathom the mysteries of NASA financial management.

In 1998, officials at the Smithsonian Institution in Washington, D.C., invited me to spend a year as the Charles A. Lindbergh Chair of Aerospace History at the National Air and Space Museum. In addition to the inspiration they created by allowing me to walk to work through the museum's gallery of flight each day, they provided much of the time necessary to research and start writing this book. I am especially grateful to Michael Neufeld, David H. DeVorkin, Dominick A. Pisano, Allan A. Needell, Tom Crouch, Anne Collins, and Valerie Neal, who supported this assignment and listened to my ideas as they took form.

My official place of employment, American University, continues to grant extensive blocks of time for research and writing, a benefit I appreciate enormously. Being a professor at a research university is one of the best jobs on earth, possibly in the galaxy insofar as we understand it. My thanks go out to the persons with whom I work, especially Cornelius M. Kerwin, Bernard H. Ross, Claire Felbinger, Walter Broadnax, William M. LeoGrande, Robert Boynton, David H. Rosenbloom, Katherine Farquhar, Richard Berendzen, Brenda J. Manley, and Cynthia Joseph, friends and colleagues all, as well as to my graduate assistant, Adam Hoffman.

I am grateful to all the people within NASA and the aerospace community who promoted the "faster, better, cheaper" initiative and gave me such a wonderful subject on which to write. Through their words and deeds, Daniel S. Goldin, Mark Albrecht, Wesley Huntress, Tony Spear, Brian K. Muirhead, Donna Shirley, Rob Manning, and many others provided the experience out of which this history is

constructed. Fellow historians and policy analysts provided insights
into the challenge of organizing complex systems and controlling
their costs. I am especially indebted to the insights of Stephen B.
Johnson, John Lonnquest, Francis T. Hoban, Charles E. Cockrell,
and Liam Sarsfield.

As usual, the professional staff at the Johns Hopkins University
Press continues to provide high standards and substantial advice. My
editor, Robert J. Brugger, his assistant, Melody Herr, and copyeditor
Maria E. denBoer pushed me to rethink and improve this work.

Finally, I wish to thank all the teachers from whom I have learned
so much, from Queen Anne High School in Seattle to Oregon State
University in Corvallis, from the University of Washington to Cor-
nell. Thank you all for launching me on such a fantastic career: Sylvia
Weinstein, Cris Kato, Ernest Charland, Charles Green, William
McClenaghan, Hugh Bone, William Harbolt, George A. Shipman,
Fremont J. Lyden, Ernest G. Miller, Morton Kroll, Clinton Rossi-
ter, A. M. Hillhouse, Harold Seidman, and, most of all, Paul Van
Riper, to whom this book is dedicated.

FASTER, BETTER, CHEAPER

THE REFORM

The high cost of spaceflight will fall soon. Engineers will design rocketships that can fly humans into space for a fraction of the cost of primitive launch vehicles. Humans will travel to Mars using technologies less expensive than those employed to dispatch the first astronauts to the moon. For years promises like these have been made by people devoted to expanding activities on this new frontier. The pledge repeats a familiar tenet of modern industrial philosophy—the belief that cost, schedule, and reliability can be simultaneously improved through advances in technology and new methods of management.

Yet spaceflight remains a hazardous technology. In the year of the agency's creation, all four missions launched by the National Aeronautics and Space Administration (NASA) failed. The management methods and technological solutions NASA officials developed to achieve reliability in the years that followed were by their very nature expensive. As humans learned how to fly themselves and their machines into space, the desire for reliability competed with thoughts of economy. Almost by definition, cheap meant unreliable.

Thirty-four years after the agency's creation, NASA executives implemented a reform designed to break the predicament this relationship posed. They proposed to develop spacecraft that were both inexpensive and reliable. They called it the "faster, better, cheaper" initiative. Through this initiative, they sought to cut costs, take greater risks, and dispatch spacecraft that actually flew. These objectives were to be achieved through changes in technology and project management. They began with small spacecraft and, when the first projects performed successfully, applied the concept to missions of increasing size. They even applied the concept to the most complicated mission of all: plans for a low-cost human expedition to Mars.

In the beginning, the initiative worked exceptionally well. Between 1992 and 1998, 10 "faster, better, cheaper" spacecraft began their work. Only one of them failed (officials canceled an eleventh project before it left Earth due to cost overruns). By historic standards, a 90 percent success rate for robotic spacecraft employing new technologies can be termed reliable.

In 1999, however, the initiative floundered. Four of the next five projects dispatched under the "faster, better, cheaper" initiative failed in flight—a clearly unacceptable outcome. The events of 1999 shook confidence in the new approach, which critics attacked as too much, too cheap, too soon. Supporters continued to cling to the original dream. The high cost of spaceflight will fall soon, they volunteered.

Achieving reliability in rockets and spacecraft is enormously difficult. Attaining reliability with economy is even harder, but it is possible, as the early history of the "faster, better, cheaper" initiative shows. The production of low-cost, reliable spacecraft can occur. But it requires the utilization of techniques that differ considerably from the methods used to achieve reliability at the beginning of the space age. It also requires careful attention to the relationship among cost, schedule, and technological complexity.

THE VISION People who believe in the promise of cheap and reliable spaceflight often point to aviation as an example of how this will occur. Airplane travel in the years following the Wright brothers' 1903 flight was risky and expensive. Constructed of wood and fabric and powered with temperamental engines, airplanes flew at substantial risk to anyone brave enough to climb onboard. Flying for all but short distances remained the province of the rich and foolhardy. All this would change soon, aviation advocates promised. Atmospheric flight would become so cheap that practically anyone could fly. Ordinary people would travel great distances across the globe in comfort. At the beginning of the twentieth century, advocates of this "winged gospel" could not explain how such changes would take place.[1] Aviation pioneers did not understand the mechanics of turbojet engines, the strength of special metal alloys, or the ability of industrial leaders to organize the construction of huge airframes. Nonetheless, aviation pioneers believed that unknown advancements in flight would soon occur. Few people believed them.

Eventually the prophecies came true. Fifty years after humans had learned to fly through the air, jumbojet technology and corporate restructuring depressed the cost of flying to a few cents per passenger mile. Advocates of spaceflight tendered a similar future for the realm above the atmosphere. In an analogous way, they said, the cost and difficulty of spaceflight would fall soon. Given that aviation pioneers took about 50 years from the date of the Wright brothers' flight to invent the necessary changes, spaceflight advocates hoped that a similar future awaited them sometime during the first decade of the twenty-first century. The "faster, better, cheaper" initiative, many hoped, would provide the experimental base from which the reinvention of spaceflight could begin.

Through the first 40 years, the cost of spaceflight remained extraordinarily high. The reusable space shuttle can deliver up to 53,000 pounds to low Earth orbit at a cost of about $8,500 per pound, or $450 million per flight. A *Titan IVB* is the most powerful nonreusable launch vehicle the United States operates. It can place 48,000 pounds in low Earth orbit at a cost of about $8,300 per pound, or $400 million every time one flies.[2] Moving beyond low Earth orbit is more expensive still. Total launch costs for the *Cassini* probe, which left Earth for the planet Saturn in 1997, topped $35,000 per pound, or $440 million.

Spacecraft costs are even higher. In 1976, two pair of *Viking* spacecraft arrived at the planet Mars. The spacecraft cost $875 million in the currency of the day or (in the value of year 2000 dollars) the equivalent of $480,000 per pound. *Viking* was not the most expensive spacecraft that the United States government had ever flown. The *Mariner 6* and *7* spacecraft that flew by Mars seven years earlier had cost $848,000 per pound to develop in year 2000 dollars. Spacecraft development costs for complex missions commonly reach $300,000 per pound.

Human spaceflight is even more expensive. The U.S. government spent $21.4 billion to place the first humans on the moon—the equivalent of $120 billion in year 2000 dollars. Given the weight of the hardware that the astronauts flew from Earth orbit to the moon, that works out to about $1.2 million per pound, including the astronauts.

Everyone in the spaceflight field knows that high costs retard the exploration of space and its exploitation for commercial purposes. Experts have tried repeatedly to reduce these costs. They have sought to reduce them by mass producing common components, such as the data recorders that all spacecraft carry. They have sought to reduce costs by reusing hardware; politicians funded NASA's space shuttle because engineers advanced the seemingly obvious proposition that a reusable spacecraft would cost less to operate than an expendable rocket that fell into the ocean after each trip. (It did not.) Engineers have experimented with various procedural reforms, such as the practice of establishing mission requirements before determining spacecraft configuration, an approach that failed to reduce costs on the International Space Station. In spite of the apparent ease with which aviation executives and leaders of other high-technology industries cut costs while improving reliability, the effort to reduce costs in the realm of spaceflight remained a history of blind alleys and dead ends.

In 1992, NASA executives tried once more. Administrator Daniel Goldin announced that NASA would develop a series of spacecraft that were "faster, better, cheaper." Goldin relied upon a seemingly well-tested philosophy to achieve this goal. He borrowed design techniques from the micro-electronics industry, asking project managers to miniaturize components as a means of reducing both spacecraft size and cost. He encouraged project managers to restructure their teams, reducing both the number of people and the time required to develop missions. In doing so, he adopted techniques previously used by executives in the private sector to downsize, or "reengineer," their corporations.[3]

Initial results seemed promising. The highly publicized Mars *Pathfinder* spacecraft and its *Sojourner* rover—the first "faster, better, cheaper" spacecraft to land on Mars—explored the Ares Vallis floodplain during the summer of 1997. The following year the low-cost *Lunar Prospector* detected evidence of water ice at the moon's north and south poles. Only two missions failed to fly. The *Lewis* Earth-observation satellite spun out of control shortly after launch, and NASA executives canceled the *Clark* satellite project when its costs ran too high.

By contrast, 1999 was a very bad year. Project managers using the "faster, better, cheaper" approach saw four failed missions that year. In March, they lost the *Wide-Field Infrared Explorer* (WIRE), a cryogenically cooled telescope, when the spacecraft's protective cover prematurely separated and drifted away. In September, the Mars *Climate Orbiter* plunged into oblivion as it approached Mars. In December, Mars *Polar Lander* and the twin *Deep Space 2* microprobes that rode with it to Mars disappeared.

In all, NASA officials attempted to fly 16 "faster, better, cheaper" missions between 1992 and January 1, 2000. The definition for inclusion in this group requires the project to have a substantially reduced cost ceiling, to be placed on a compressed development schedule, and to have someone with executive responsibility in NASA designate it as a "faster, better, cheaper" project. All of the spacecraft were small and most of them tested new technologies. Of the 16 projects undertaken, six had failed as of January 1, 2000. The resulting success rate, a paltry 63 percent, was significantly below spacecraft norms and well below the expectations for reliability that advocates placed on the initiative.

THE DEBATE Assessment of the "faster, better, cheaper" initiative comprises a subset of a larger debate between people who believe that

1   THE FIRST "FASTER, BETTER, CHEAPER" PROJECTS

Discovery Program

NEAR (Near Earth Asteroid Rendezvous). Launched February 17, 1996. Completed a flyby of the asteroid Mathilde in June 1997; failed to complete an engine burn in December 1998, preparatory to orbiting asteroid Eros; spacecraft systems recovered, orbital insertion achieved in February 2000.

Mars *Pathfinder.* Launched December 4, 1996. Landed on Mars July 4, 1997; dispatched *Sojourner* rover and conducted studies of Ares Vallis floodplain.

Lunar Prospector. Launched January 6, 1998. Orbited the moon; discovered evidence of water ice at the moon's north and south poles; mission completed July 1999.

Stardust. Launched February, 7, 1999; scheduled to encounter comet Wild 2 in 2004 and return samples of comet material to Earth in 2006.

New Millennium Program

Deep Space 1. Launched October 24, 1998, 3 months late. Flight-tested 12 new technologies, including an ion propulsion engine. Encountered asteroid Braille in July 1999 and, if the spacecraft continues to function, will fly by a comet and a comet-like body in 2001.

Deep Space 2. Launched with the Mars *Polar Lander* on January 3, 1999. Contained two micro-probes designed to penetrate the Martian subsurface and search for evidence of water ice. Both probes disappeared during the landing phase.

Mars Surveyor Program

Mars *Global Surveyor.* Launched November 7, 1996. Reached Mars September 11, 1997; difficulties with aerobraking maneuvers delayed entry into final orbit; undertook extensive mapping activities.

Mars *Climate Orbiter.* Launched December 11, 1998. Scheduled to arrive September 1999, and provide detailed information about the surface and climate of Mars. Lost when it passed too close to Mars, a result of miscommunication about the units of measurement used to calculate the position of the spacecraft.

Mars *Polar Lander.* Launched January 3, 1999. Scheduled to land on December 3, 1999, and investigate an area near the south pole of Mars. Lost during the entry-landing phase.

Small Explorer Program (SMEX)

Solar, Anomalous, and Magnetospheric Particle Explorer (SAMPLEX).
Launched into Earth orbit July 3, 1992. Conducted wide-ranging investigations of local interstellar matter and solar material.

Fast Auroral Snapshot Explorer (FAST). Launched into Earth orbit August 21, 1996, 2 years late due to problem with the *Pegasus XL* launch vehicle. Conducted extensive investigations of physical processes that produce aurora.

Submillimeter Wave Astronomy Satellite (SWAS). Launched December 4, 1998, 3.5 years late due to problems with the *Pegasus XL* launch vehicle. Conducted telescope investigations of the composition of dense interstellar clouds where new stars form.

Transition Region and Coronal Explorer (TRACE). Launched April 2, 1998, into a sun-synchronous orbit (a polar orbit in which the satellite is always over the dawn-dusk line of the Earth). Employed a special telescope to take spectacular high-resolution photographs and conduct investigations of the sun's atmosphere.

Wide-Field Infrared Explorer (WIRE). Launched March 4, 1999. Contained a cryogenically cooled telescope and arrays of infrared detectors to study the evolution of galaxies. Declared a total loss after the telescope's protective cover was prematurely ejected shortly after launch, causing the frozen hydrogen needed to cool the telescope to vent into space.

Small Satellite Technology Initiative

Lewis. Launched into Earth orbit August 22, 1997, one year late due to problems with the Lockheed Martin *Athena* launch vehicle. Contained advanced Earth-sensing instruments. Spun out of control 4 days after launch; total loss.

Clark Earth Observing Satellite. Designed to carry a variety of instruments to study the Earth and sun, including a very-high-resolution camera for taking stereo images of the Earth. Project terminated February 1998, due to cost overruns and schedule delays.

high-technology undertakings lead to "normal accidents" and those who think that humans have discovered techniques for creating highly reliable organizations. Since industrialized societies are depending on a future of high technology, the outcome of the debate will influence the course of civilization in the twenty-first century.

Complicating the debate is a third point of view, the so-called New Public Management, also known as "reinventing government." This movement began in the 1980s as a result of dissatisfaction with old-line government bureaucracies and the inability of civil servants to respond to rapid technological change. The New Public Management encompasses a variety of techniques (reengineering, downsizing, outsourcing, privatization, and competition) designed to create institutions delivering public services that are cheaper to operate, more responsive to the "customers" they serve, and better able to incorporate new technologies into their production processes.[4] For many government agencies, including NASA, the New Public Management represented an effort to economize.

Some experts believe that risky technologies are by nature unreliable and will eventually fail. Other people believe that humans have perfected methods for overcoming failure in the management of complex technologies. But they are suspicious of efforts to economize. Still others promote the New Public Management. They would like to achieve both reliability and economy.[5]

"Faster, better, cheaper" provided a major test of these conflicting points of view. During the early period of mission success, it was the poster child for the New Public Management and the promises it made. The significance of the "faster, better, cheaper" initiative went well beyond the space program to complex endeavors of many forms.

Advocates of "faster, better, cheaper" insist that the concept is sound, in spite of the unanticipated failure rate. They point to errors that would not have occurred if managers had followed appropriate procedures. They hope that such errors will disappear as people become more familiar with the approach. To its advocates this approach is sound in theory but difficult to implement in practice.

Critics of the approach do not agree. They acknowledge that engineers can develop small, simple spacecraft with limited capabilities at low cost. Small spacecraft have a long and noble history in the aerospace field. They can be produced quickly and cheaply. They have worked well, especially for simple missions. According to the critics, however, simple spacecraft are not appropriate for big tasks. With few exceptions, they cannot explore planets or observe the universe or protect humans in the radiation-soaked vacuum of space. As mission complexity intrudes, the advantages of small spacecraft disappear. For complex missions, the procedures for developing small spacecraft invite failure, especially when project directors attempt to cut development time. In simple terms, "faster" conspires with "cheaper" to eliminate "better" on complicated missions. The relationship has been demonstrated empirically by space analysts like Liam Sarsfield and David Bearden.[6]

An old tenet widely accepted by aerospace workers holds that cost, schedule, and reliability are interrelated in such a way that attempts to improve one adversely affect the others. From this point of view, reduced schedules inevitably drive up costs (when the level of reliability is held constant). The history of the Apollo moon race, the crash program set up to place Americans on the moon, illustrates this point of view. The need to launch the expedition within eight years—and bring the astronauts back alive—encouraged costs to soar. Conversely, efforts to cut cost often lengthen completion schedules. The International Space Station, originally set for completion in 10 years, took more than twice that long due to efforts to restrain cost growth through frequent redesign. Risk and reliability introduce a third dimension. Efforts to reduce risk drive up costs and lengthen schedules. The latter relationship is demonstrated by the reusable space shuttle, in which efforts to maintain safety on a risky technology spoiled the twin goals of cheap and frequent spaceflight.

The apparent relationship among cost, schedule, and reliability creates what many in the aerospace field call a "pick two" philosophy. According to this point of view, aerospace workers can improve two of the three terms in the "faster, better, cheaper" equation, but not three. By the laws of physics, according to this point of view, faster and cheaper cannot simultaneously be better.

RESOLVING THE DEBATE The history of the "faster, better, cheaper" initiative, and the evidence that has been collected to date, suggests the following conclusions. Engineers and other experts can reduce the cost of spaceflight and the time necessary to prepare missions for flight. Moreover, they can do so without significant loss of reliability (90 percent mission success rates for robotic spacecraft). They can also do so with only modest reductions in spacecraft capability. The "better" in "faster, better, cheaper" means that spacecraft deliver more capability for the dollars invested (more "bang" for the "buck") than large, expensive spacecraft. "Better" means increases in relative, not absolute, capability.

This is accomplished in two ways. First, project workers use micro-technologies to reduce the size and weight of the spacecraft they produce. Less weight reduces cost, since cost per pound tends to be relatively fixed for any given class of spacecraft. Less weight frequently reduces spacecraft complexity, especially for those projects that seek simplicity through miniaturization. Reduced complexity also saves money, since simple projects cost less to manage.

Second, project workers utilize a simpler, less expensive system for managing the project. Traditionally, spacecraft workers have utilized systems engineering (or systems management) as the principal means for fighting failure in risky systems. Systems engineering is a technique developed within the American aerospace industry that provides a formal process for coordinating the parts of a rocket or spacecraft that must work in concert in order for the system to fly. Systems management is formal, elaborate, and expensive. Workers on "faster, better, cheaper" projects forego many of the aspects of formal systems management. Instead, they rely upon techniques that, in a phrase, substitute teamwork for paperwork.

Teamwork techniques are cheaper and require fewer people on the project payroll. Such techniques are workable for smaller, simpler projects in which workers can resolve reliability problems through face-to-face communications instead of formal paperwork reviews. By reducing the size and complexity of spacecraft, advocates of "faster, better, cheaper" increase the likelihood that team-based management techniques can be employed to assure reliability.

So what went wrong in 1999? Analysis suggests two principal causes for the failures that afflicted missions that year. The first concerns what can be called the Bearden rule, named after one of the first analysts to empirically examine the relationship among cost, schedule, complexity, and mission success. According to David Bearden, complexity increases cost and development time on successful missions. The relationship is linear for schedule and exponential for cost.

When aerospace executives depress cost or schedule below the level required by mission complexity, they invite failure. "When examined after the fact," Bearden says, "loss or impaired performance is often found to be the result of mismanagement or miscommunication," difficulties traceable to budget ceilings or schedule constraints that are too tight for the project at hand. Under these constraints, managers lack "sufficient resources to test, simulate or review work and processes in a thorough manner."[7]

Second, project managers often fail to institute the full scope of techniques necessary to assure that teamwork will control reliability. Business managers and government bureaucrats accustomed to Cold War largess cling to traditional management practices that cost a great deal. When faced with budgetary constraints, managers abandon features associated with systems management yet resist the teamwork required by low-cost techniques. As a consequence, managers fail to institute either of the principal methods available for managing risk (systems management or teamwork).

This is largely a cultural issue. It grows out of resistance within the aerospace field to changes in organizational practices required by low-cost spacecraft.

The future of "faster, better, cheaper" will be determined by the ability of its advocates to apply the approach to projects of ever-increasing size. Even critics of the approach acknowledge that the technique can be applied to small, relatively uncomplicated spacecraft. Agreement breaks down as complexity increases. Advocates of the approach want to apply it to larger, more complex endeavors—new space observatories, rocketships designed to replace the aging and expensive fleet of space shuttles, and plans for a low-cost human expedition to Mars. Critics doubt that this can be done.

Analysis suggests that teamwork principles and micro-technologies cannot be utilized with success for planetary expeditions and shuttle-class rocketships with humans onboard at the cost levels desired by advocates of the approach. The missions are simply too complex. Medium-size missions such as space observatories may prove amenable to the approach, but the really big missions will not.

For people who want to believe in the spacefaring dream, with its images of exotic spacecraft and planetary colonization, the promise of cheap and easy activity is hard to resist. Casual observers of the aviation and personal computer industries want to imagine a similar future for space travel. Political leaders who view high technology as the means for growing economies and solving problems such as global warming naturally support the philosophy that difficult and expensive barriers will not block the way.

Wanting to believe is not commensurate with being there. Cheap and easy spaceflight is attainable, but harder than it looks. It requires management reforms that are inconsistent with the traditional culture by which humans first learned how to venture into this final frontier and it requires the mastery of new and unfamiliar technologies.

THE NATURE
OF THE CHALLENGE

Preventing failure when dealing with complex technologies is a challenging task. The nuclear core at the Three Mile Island power plant nearly melted down in 1979 when operators lost control of the cooling process. The French *Concorde* supersonic transport caught fire, crashed, and burned in 2000 when fragments from a ruptured tire pierced the fuel system on the left wing. Seven *Challenger* astronauts died when hot gasses breached o-rings on one of their solid rocket boosters. Three American astronauts perished when a fire broke out in the all-oxygen atmosphere of their *Apollo 204* test capsule.

Automated and robotic systems are especially hard to build. The first six automated *Ranger* spacecraft sent to take close-up pictures of the moon failed to complete their tasks. Astronomers could not focus the *Hubble Space Telescope* because technicians improperly ground its primary mirror. The *Galileo* spacecraft to Jupiter could not unfurl its high-gain antenna.

Spacecraft and other high-technology systems can fail in hundreds of ways. The people who design and build them fight failure every day. They build in redundancy, so that secondary units take over when primary ones fail. They test components extensively, especially those bound for the harsh conditions of space. They identify components whose failure would jeopardize the mission and install sensors that monitor their status and communicate that information to operators and pilots. They build in safety features that throw systems into safe modes when difficulties occur. They use systems management techniques and risk management studies to prevent failure.

In 1984 the sociologist Charles Perrow wrote a book analyzing the reliability challenge in societies that utilize risky technologies. In *Normal Accidents,* Perrow argued that very complex systems possess characteristics that inherently cause them to fail. Accidents are normal, Perrow advised, given the architecture of modern technology.[1]

Perrow was especially concerned with what engineers call interactive failures. Most engineers can identify single points of weakness in complex systems and counteract them. Difficulties arise when two or more components interact in unexpected ways. An interactive failure doomed the flight of the Mars *Polar Lander* in 1999, one of NASA's most ambitious "faster, better, cheaper" projects.

MARS *POLAR LANDER* NASA flight engineers planned to set the robotic Mars *Polar Lander* on the surface of Mars on December 3, 1999. The lander carried a parachute and descent engine designed to permit a soft landing. To save money, the people who designed the lander did not include telemetry that could provide data on the status of the lander as it dropped to the ground. Put simply, Mars *Polar Lander* had to land alone, out of communication with humans back on Earth. As the spacecraft began its descent, the communication blackout occurred as planned. Flight controllers never heard from the spacecraft again.

A special accident review board concluded that the lander had probably crashed when its descent engine prematurely shut down about 130 feet above the ground. The descent engine should have shut down when the lander legs touched the surface of Mars. Each of the three landing legs contained a magnetic sensor designed to generate voltage when the legs struck the surface. Spacecraft designers programmed onboard software to recognize the voltage signal and shut down the engine. The engine had to shut down within 50 micro-seconds after touchdown. If it kept burning, it would tip over the spacecraft. The sensors had to work quickly in order to provide a safe landing.

Unfortunately, the sensors also tended to generate voltage when the spacecraft deployed its three landing legs. This occurred at an altitude of about 5,000 feet, about 4 minutes after entry, while the lander was still attached to its parachute. Spacecraft designers recognized this irregularity and programmed the flight software to ignore it. The program logic was complicated—it used six variables both to eliminate false signals from a working sensor and to avoid the equally disastrous possibility of no signal from a failed sensor. In writing the program, however, the designers made an error. They wrote a line of code that caused a premature shutdown signal at 5,000 feet to be recognized as a true signal at an altitude of 130 feet when the radar altimeter signaled the computer to prepare for engine shutdown.

This anomaly could have been caught during spacecraft tests on Earth. It was not. A leg deployment test was performed on June 4, 1998, with the flight software running. In that test, the leg sensors did not work, failing to detect any touchdown at all. Designers discovered the cause, an error in a wiring diagram that prompted workers to incorrectly wire the sensors. Workers corrected the problem and technicians repeated the test. In doing so, however, they made a serious error. They repeated only the touchdown phase, not the leg deployment sequence. At the simulated moment of touchdown, the sensors worked. Had the technicians simulated leg deployment, they might have discovered the spurious signal. Focusing on the equally troublesome problem of no shutdown, they failed to detect the disastrous probability of a premature one.

In hindsight, the error seems obvious. To people working on the spacecraft before it flew, it was not so easy to see. Spacecraft designers were concerned with many interactions that could doom the mission, not just the behavior of touchdown sensors. System engineers believed that they had solved the premature shutdown problem by instructing the computer to ignore spurious signals prior to what they called the 40 meter altitude check. Mechanical design engineers who constructed the landing legs understood that the legs tended to generate spurious signals, but did not communicate exactly how that might occur to system engineers. As a consequence, the people who designed the spacecraft failed to write a requirement that might have corrected the problem during unit- or system-level testing. In the words of the investigating team, "there was no software requirement to clear spurious signals prior to using the sensor information to determine that landing had occurred."[2]

The loss of Mars *Polar Lander* was a classic "normal accident," as people like Perrow have described them. It required the interaction of three separate systems—landing legs, touchdown sensors, and a software program—in ways that spacecraft designers had failed to anticipate. The failure took place quickly, in about 60 seconds, the period between landing leg deployment and the 40 meter altitude check. At that altitude, spacecraft engines shut down and the robotic spacecraft crashed into the ground.

The accident investigation team could not confirm this anomaly as the only possible cause. Something else could have gone wrong in the 5 minutes between entry and the altitude check. With no spacecraft telemetry during entry, it was hard to tell. Careful investigation convinced members of the investigating team that the spacecraft would have crashed once it reached the altitude check, assuming that no earlier catastrophic events occurred.

Traditional technologies tend to be linear in their operation, with one event following another in a logical sequence. They operate like a railroad train, with all the cars in a row. Errors occur in linear processes, but they happen in ways that are predictable and thus easier to control.

Traditional technologies also tend to be loosely coupled. In loosely coupled systems, substantial gaps exist between the advent of problems and catastrophic failure, in the form of slack or buffers between steps in sequence. Like a slow-spreading disease, errors in loosely coupled systems do not reverberate throughout the whole operation. They remain localized in such a fashion that operators can find sufficient time to correct them through technological fixes or revised procedures.

In the world of high technology, complexity intrudes. Complex systems like spacecraft are more interactive, with components related to one another in intricate ways. They contain elaborate feedback loops; controls are numerous and have a tendency to interact. The systems, moreover, tend to be tightly coupled, with the result that simple failures rapidly cascade into full-scale catastrophes. By definition, complex systems are characterized by interactive relationships and tight coupling. The combination of these two features creates circumstances under which big accidents, in Perrow's view, inevitably occur.

Spaceflight engineers fight this tendency in a number of ways. They increase redundancy. They do more testing. They install more sensors. They adopt more safety features. They write more precise requirements. They expand their use of formal systems management, in which changes in a single component are analyzed for the effects they might have on other subsystems. From Perrow's point of view, however, all such efforts are self-defeating. They increase system complexity, and as technical complexity moves up the scale, so does the tendency to fail.[3] To avoid failure, engineers set upon this course of action must add more redundancy, tests, sensors, safety features, requirements, and systems management. The result is a never-ending spiral of increasing costs and risk avoidance procedures, often accompanied by a lengthening schedule.

Controlling risk in this manner is like trying to push a spacecraft toward the speed of light. An object approaching light speed increases in mass, requiring ever-greater inputs of energy to accelerate it. Physicists point out that the exponential nature of this relationship would require a spaceship captain to apply an infinite amount of energy to propel an object at light speed. Faster-than-light transportation is fun as fantasy. Hopes of attaining such speeds in nature, so far as scientists understand it, are illusory. In the same manner, risk elimination on very complex spacecraft requires an infinite amount of work and also remains an elusive goal.

The traditional approach to this problem, especially for human spacecraft, is to use all the redundancy, safety features, testing, telemetry, and systems management that money can buy. This "can't fail" approach to spaceflight was very popular during the early decades of space exploration, when aerospace technology was immature and Cold War anxiety encouraged large outlays of cash. It guided the progress of Project Apollo, which in spite of the notion that it could not fail, resulted in the death of three astronauts during a launch pad fire and the near-demise of three more astronauts on the flight of *Apollo 13*.

THE LOSS OF MARS *OBSERVER* Results from the "can't fail" approach discouraged NASA executives in a number of ways. Mission costs continued to grow, almost uncontrollably. Concurrently, spacecraft continued to fail. Loss of the Mars *Observer* spacecraft in 1993 provided an especially bitter episode.

In the minds of the people who conceived it, the spacecraft should have been the first in a series of lower-cost spacecraft to visit Mars. But it was also the first machine to return to the red planet in nearly 20 years, and scientists argued for instruments that a simple robot could not handle. The mission grew in cost and complexity, and advanced in age to the point that 11 years passed before the spacecraft was ready for launch, dated from its point of conception in 1981.

Humans stuffed the spacecraft with instruments. As the size of the spacecraft grew, so did its budget, which increased from $250 million to more than $800 million. Increased spending and a lengthened development schedule should have bought a more reliable spacecraft. It did not. As the spacecraft approached Mars on August 21, 1993, it disappeared. Members of a special investigating team traced the failure to a problem with the 14 minute sequence for pressurizing the spacecraft's fuel tanks prior to the orbital insertion engine burn, compounded by shortcomings with the spacecraft's radio communication system, an interactive relationship.

Loss of the Mars *Observer* helped build support for the "faster, better, cheaper" initiative. For the price of a single, large mission, aeronautical engineers could launch many small ones. Mars *Polar Lander* cost about $165 million; *Lunar Prospector* consumed an inconsiderable $64 million. Advocates of "faster, better, cheaper" paid for their first 16 missions with less money than expended on one large planetary expedition. The first 16 missions in inflation-adjusted dollars cost less than the *Cassini* mission launched on a long voyage to Saturn in 1997. For the price of one traditional planetary venture, NASA executives bought five missions to Mars, one mission to the moon, three space telescopes, two comet and asteroid rendezvous, four Earth-orbiting satellites, and one ion propulsion test vehicle.

Strictly in terms of cost, the "faster, better, cheaper" approach produced spacecraft that delivered more. The trick, of course, is to produce spacecraft that are reliable as well. People promoting the initiative knew that they would have to take greater risks. They knew that they might lose one or two missions. They also knew that they could not use traditional methods for fighting failure, methods that led inexorably to higher costs. Partisans of the new approach experimented with methods for reducing cost and failure that were fundamentally different from those used to combat failure on more complex missions.

THE CHALLENGE OF HIGH RELIABILITY According to skeptics like Perrow, accidents are inevitable when societies employ risky technologies. In 1984, a group of social scientists based at the University of California at Berkeley began to study people whose institutions, although confronted by complexity, had achieved very high levels of reliability. The scientists studied the Federal Aviation air traffic control system, the U.S. Navy fleet of nuclear-powered aircraft carriers, and the nuclear power generating plant at Diablo Canyon, California. As this line of inquiry progressed, other researchers examined NASA's civil space program, which by that time had slipped into a period of unreliability initiated by the loss of the space shuttle *Challenger*.[4]

High-reliability organizations have the characteristic of "working in practice but not in theory."[5] Perrow's theory predicts that humans should not be able to create nearly error-free institutions managing risky technologies in natural surroundings. The existence of such institutions, including NASA during its moon landing years, contradicts the theory.

Workers in high-reliability organizations take a number of steps that apparently allow them to manage risky technologies without committing extensive errors. They adopt a "culture of reliability," in which the norm of safe operations is widely shared. The organizations are simultaneously centralized and decentralized, with strong hierarchy and standard operating procedures coexisting with situations in which workers possess considerable technical discretion. People in such institutions utilize different organizational structures for different tasks. They make extensive use of redundancy, in both mechanical systems and operating procedures. And they engage in a process of constant learning, frequently through risk-taking, trial, and the analysis of error.[6]

The Berkeley group focused their research on people in institutions like nuclear power generating plants who operated risky technologies. Space exploration is different in one significant regard. People engaged in space exploration spend considerable time designing and testing new technologies before they attempt to fly. To achieve reliability during what is called the design and development phase, NASA employees emphasize a special feature commonly found in high-reliability organizations. They simultaneously centralize and decentralize.

To achieve centralization, NASA employees have traditionally relied on systems management. In his history of organizational practices in the U.S. space program, Stephen Johnson calls systems management "the secret of Apollo."[7] Systems management provides highly structured, rigid procedures for tracking design changes and wedding them to cost, schedule, and the proper functioning of other subsystems. To achieve decentralization, NASA executives promote features that tend to lodge discretion with experts in the field—features like in-house technical capability, hands-on activity, extensive testing, risk-taking, and constant learning. Systems management creates the glue that binds project organizations together; technical discretion creates expertise. The complicated system enabled humans to fly to the moon and robots to investigate other bodies in the solar system.[8]

People familiar with the confluence of systems management and technical discretion view cost-cutting suspiciously. The features are designed to produce reliability, not save money. Writing generally about high-reliability organizations, Martin Landau has called the economy doctrine "the most pernicious constraint ever laid on a public organization." Even if one strives for economy, says Landau, "it makes no sense to strive for it until one has a measure of effectiveness."[9] Another high-reliability researcher, Paul Schulman, worries that pressures to economize will reduce the "slack" or "buffers" that workers in such institutions need to resolve unexpected problems as they arise.[10]

METHODS Proponents of the "faster, better, cheaper" initiative sought to achieve economy without significant reductions in reliability. They attempted to do this in a number of ways. To achieve reliability and economy simultaneously, they developed spacecraft simpler than the missions that had preceded them. The primary route to simplicity runs through size reduction, a development made possible by microtechnology. Smaller spacecraft cost less to construct. Reliability problems on small spacecraft can be resolved through teamwork, which reduces both the scale and the cost of management. This was the basic approach that characterized the initiative.

In many ways, these people sought to return to the values that had characterized NASA's organizational culture in the first decade of spaceflight before the arrival of systems management. In the early years, small teams tested, built, and operated spacecraft without much outside interference. Leaders of modern teams attempted to do it again, this time with the advantages produced by 35 years of spaceflight experience and the advent of more sophisticated, small-scale technologies.

Traditionally, spaceflight managers have fought risk with complexity, in both the mechanical systems they design and the management methods they use to coordinate their work. Rather than chase reliability with complexity, champions of "faster, better, cheaper" pursued simplicity. Borrowing lessons from the history of small spacecraft, they reduced the size of the machines being designed and the size of teams needed to manage the missions. In theory, a small project team can attack problems of reliability without resorting to the elaborate systems management techniques used to combat interactive failures. People on small teams can solve reliability issues through informal communication, a consequence of the smaller number of workers with whom they must deal.

Spacecraft can be made smaller through micro-technologies. The technological maturity of components like computers and cameras allows project teams to gain the advantages that accrue from smaller size. Smaller, lighter spacecraft are easier to operate; they are cheaper to produce. They can be launched on smaller rockets, thereby promoting cheaper access to space. Smaller spacecraft are less complex and thus less susceptible to interactive failures. Being smaller, they may have less capability than larger machines, but they may deliver more for their size than weightier models.

To cut size, leaders of "faster, better, cheaper" projects also reduce spacecraft capability. The first set of "faster, better, cheaper" spacecraft did not possess the capability of *Cassini*-class machines. Reduced capability does not mean that a mission is automatically worse. A mission with one-half the capability will be "better" if it performs that mission at one-tenth the price. It delivers more for its size than weightier models. In a similar fashion, small size need not diminish capability proportionately. New cameras are smaller than ones used on earlier spacecraft, but produce better images due to advances in digital photography.

"Faster, better, cheaper" teams utilize fewer people and employ them for shorter periods. Compared to traditional missions, team leaders may employ one-third of the people on a development schedule half as long. They may employ a small operations group to guide the spacecraft after launch. Small teams cut overall costs dramatically and allow less formal management methods.

With fewer people, project managers cannot perform the elaborate systems management studies that have been used traditionally as a guard against catastrophe. They do not have enough people to complete all of the paperwork required by formal systems management. This can be an advantage when project groups are small, because it forces workers to substitute teamwork for paperwork. They have no other choice.

Traditionally, spacecraft engineers have employed formal systems management as their primary tool for controlling risk. Experience beginning with the very first missions to go into space suggests that engineers who ignore systems management will more likely fail. People working on "faster, better, cheaper" projects use some elements of systems management, but it is not the critical factor for keeping risk, cost, and schedule low. Teamwork is.

In theory, a small team can develop the capacity to recognize risks, solve problems, and learn from errors without resorting to formal procedures. Members of such teams create what is called a "self-learning" organization, enhancing the capacity for both risk-taking and high reliability. To do this, however, members must work hard to build the team. Where teamwork is weak, problems can spin out of control. Having cut expenses and foregone formal systems management, groups without the capacity for teamwork lack the means to control risk. They have no cushion on which to fall when problems occur. Teamwork principles are a substitute for formal systems management, but there is not much margin for error.

"Faster, better, cheaper" leaders adopted a number of techniques designed to promote teamwork on their projects. The techniques were identified by project leaders and confirmed in NASA reports on the approach. The main techniques are summarized below. In practice, their application varied from project to project, and in some projects, a few were not applied at all. In general, however, the techniques formed the basis for the effort within NASA to promote teamwork while restraining costs.

1. Cost goals. Project leaders accept cost containment (and schedule control) as a major goal. They communicate this goal to team members, who rank it as high in importance as the scientific and technical objectives of the project.

2. Project scale. The spacecraft is small and the project team that develops and flies it is very small.

3. Experienced and inexperienced personnel. Lacking formal safeguards, project leaders recruit experienced personnel who can recognize risks and resolve technical problems, and mix them with a larger number of inexperienced personnel.

4. Technical discretion. Team members are allowed to control their own work, which includes the authority to make design changes and supervise contractors without outside interference.

5. Protection. Team members are protected against red tape, annual budget caps, excessive outside review, external micromanagement, and other outside forces that threaten to limit their discretion.

6. Stable funding. Within the total program cost cap, team leaders have the ability to spend funds at the most appropriate point in time. They receive funds when funds are needed.

7. Co-location. Divided work packages and multicenter projects seriously hamper team effectiveness. Components of the spacecraft may be developed at different locations, but the central management team is located at one place.

8. Multitasking. Multitasking is the process of moving team members from one job to another as the project matures, increasing the overall sense of responsibility and organizational memory.

9. Hands-on activity. Team members learn about the spacecraft

by working with actual hardware. They build and test the spacecraft themselves.

10. Testing. Extensive testing, along with hands-on activity, allows team members to become intimately familiar with the intricacies of the spacecraft. It also reduces risk.

11. Seamless management. Seamless management is the practice of using the same people on the project team from design through operations even as the nature of the work changes. This reinforces the principle of multitasking.

12. Peer review. As a partial substitute for the checks lost by foregoing systems management, project workers present their plan to groups of their peers.

13. Cancellation. Top executives are prepared to cancel any project where costs or schedule spiral out of control, and team members are told that this will happen.

14. Risk-taking. Team members are encouraged to be creative and take calculated risks (but told not to fail).

15. Risk management. Although project leaders do not emphasize formal systems management, they practice risk management techniques.

The location of the central project team does not matter. Industry groups operating under government contracts managed some projects; others were carried out at NASA field centers. The overall goal—a strong, cohesive, and independent team—remains the same no matter where the project is located.

EXPLAINING FAILURE So why did so many missions fail? The answer, according to people who have analyzed the problem, lies in the relationship among cost, schedule, and complexity. In short, proponents of the approach created failure when they reduced cost and schedule faster than they lessened complexity. They also invited failure by neglecting to adopt all of the teamwork principles necessary to compensate for the reduction in systems management.

Reducing spacecraft complexity is a major tenet of the approach. In their studies of small spacecraft, Sarsfield and Bearden identify a number of factors that affect spacecraft complexity. Bearden uses 21 factors to measure relative complexity. Some of the more important factors are summarized below:

▲ Spacecraft mass at launch
▲ Electric power requirements
▲ Distance that the spacecraft will travel from Earth
▲ Expected life of the spacecraft
▲ Pointing accuracy
▲ Methods of propulsion and stabilization
▲ Maximum downlink data rate and solid-state recorder memory
▲ Number of instruments

Sarsfield and Bearden demonstrated a direct relationship between cost and schedule and complexity. Sarsfield reviewed 12 NASA spacecraft, including 10 of those covered in this study, plus one launched by the Department of Defense. All of the spacecraft were small. Complexity multiplied by the weight of each spacecraft (dry mass, without propellants) accurately predicted their cost. "The cost of small spacecraft correlates well with dry mass," he observed, "when corrected for complexity."[11] Sarsfield's study provided empirical evidence for the commonly held belief that complexity affects cost.

When government executives attempt to reduce cost (or schedule) below the predicted value, they interfere with the cost-complexity relationship. The result, not unexpectedly, is higher risk and less reliability. Using the complexity criteria developed by Sarsfield, David Bearden investigated the consequences of excessive cost and schedule reduction. Bearden examined 21 NASA spacecraft, including 14 of the 16 projects covered in this study. He began by creating baselines that expressed the relationships among cost, schedule, and complexity based on an analysis of 22 small satellites, previously developed, all of which flew successfully. The baselines revealed a linear relationship between complexity and schedule (more complex satellites required longer development times) and an exponential relationship between complexity and cost (the cost of complex satellites rose along a steadily ascending curve).

When Bearden applied the baselines to NASA's low-cost initiative, patterns emerged. Roughly half of the projects that Bearden studied failed or were impaired in some fashion. They clustered together at the high end of his complexity scale underneath the line that predicted the desired development time. The development schedule was too short given the complexity of the spacecraft. Project schedule strongly predicted mission reliability.

Bearden's findings on cost were more perplexing. Low-cost missions of the lowest complexity, clustered along the baseline defining adequate spending, did very well. As projects moved up the cost/complexity curve, failures occurred. It did not matter whether the more complex "faster, better, cheaper" projects sat above or below the adequate-spending line. All of the more complex projects failed or were impaired, regardless of whether their managers spent too little or too much.

An analysis of Bearden's findings, taken in conjunction with a review of the accident reports for failed missions, supports two conclusions. First, increased spending does not compensate for projects that managers develop too rapidly. A project developed too fast and too cheaply will almost certainly develop problems of reliability, but so will one that is simply developed too fast. According to Bearden, schedule is a more powerful predictor of reliability than the amount of money spent on the project.

Second, many of the project managers who developed more complex "faster, better, cheaper" spacecraft failed to follow the teamwork principles associated with the initiative. In combination with short development schedules, this was a certain predictor of reliability problems once the spacecraft flew.

These findings are illustrated by the outcome of two of the 1999 missions that project managers sent to Mars: *Deep Space 2* and Mars

Climate Orbiter. Managers dispatched a pair of *Deep Space 2* micro-probes to Mars in the same container as Mars *Polar Lander.* They separated from the lander as the entry sequence began. The two micro-probes were designed to fall through the thin Martian atmosphere, unaided by parachutes or any landing devices, and smash into the Martian soil. Spacecraft engineers designed each micro-probe so that the force of impact would propel a tiny instrument about 2 feet deep into the Martian soil, where it would search for water. The instrument, called a forebody, would communicate its findings to the remainder of the micro-probe resting on the surface of the planet, which in turn would radio the results to a satellite orbiting Mars and back to Earth. Both parts of the micro-probe had to survive the force of a landing calculated to exceed 400 miles per hour.

NASA flight engineers received no signals from either of the 5 pound probes. The investigating team blamed the loss on a development schedule that was simply too short. "It is clear," investigating team members concluded, "that the microprobes were not adequately tested and were not ready for launch."[12]

The fated Mars *Climate Orbiter* illustrates the second issue. NASA employees have a long tradition that favors in-house technical capability and hands-on work. The tradition predates the agency's creation, and can be found in the organizations out of which NASA was formed. From the beginning, NASA employees developed a second tradition, one of contracting out work to business firms. Roughly nine-tenths of the money that Congress appropriates for NASA programs goes to NASA contractors. The two traditions clash with each other, producing projects simultaneously managed by both government workers and industry employees.

Historically, spacecraft managers have resolved the conflict between in-house control and contracting out by employing systems management. Systems management provided the solution to the problems of coordination that inevitably arose, welding otherwise disorganized groups into what executives called their "government-industry team." The so-called team was in fact a collection of teams joined together through formal management techniques.

The "faster, better, cheaper" doctrine alters that approach. In their quest for cost reduction, its partisans deemphasize the use of systems management. This requires managers to adopt compensating techniques. One of the most common is co-location, the creation of a single project team at one site. This technique, however, clashes with the old traditions associated with contracting out. The result, in the case of Mars *Climate Orbiter*, was a project run under old traditions of split control with none of the formal safeguards offered by systems management.

The spacecraft was built by a group of employees at the Lockheed Martin Astronautics complex in Denver, Colorado. They were responsible for spacecraft design, development, and systems integration. They were supervised by a separate group of employees at NASA's Jet Propulsion Laboratory (JPL) in Pasadena, California. JPL employees were responsible for instrument development, systems engineering, navigation, mission operations, and overall project management. People working on spacecraft navigation at JPL were not involved in spacecraft design. They did not attend the critical design review meetings held by the people building the spacecraft. The JPL navigators assumed that Mars *Climate Orbiter* would operate much the same as Mars *Global Surveyor*, another "faster, better, cheaper" mission already under way.

Navigators calculate spacecraft location using formulas that predict the way in which thruster firings alter spacecraft velocity. They check those solutions by observing Doppler signatures from traveling spacecraft. For Mars *Climate Orbiter*, results from the two sets of calculations did not agree. No one resolved the discrepancy. On September 23, 1999, Mars *Climate Orbiter* swept behind the planet and disappeared. Six days later navigation experts discovered that the velocity changes reported by spacecraft engineers were low by a factor of 4.45. One pound of force, using English measurements, converts to 4.45 Newtons. As unbelievable as this sounds, the space-craft engineering group in Denver had been calculating velocity changes using English measurements while the navigation group in California was working in metrics. Divided as they were, the two groups failed to recognize the discrepancy. As a result of this error, the spacecraft swept too low into the Martian atmosphere and disappeared.

The "faster, better, cheaper" approach can overcome problems of reliability. To do so, however, requires close adherence to a number of principles. The designing team must reduce spacecraft size and complexity through micro-technology. Supporters must fund the project at a level commensurate with its complexity. They must provide an adequate development schedule, again based on mission complexity. And they must rigorously utilize team-based management techniques.

Early experience with the approach suggests that the latter requirement is particularly difficult to achieve. It entails, in the words of the group investigating the loss of Mars *Climate Orbiter*, "a new NASA culture and new methods of managing projects."[13] Culture shifts do not occur easily.

COST CONTROL

The challenges of spaceflight conspire to defeat efforts at cost control. The harshness of space, the great distances that spacecraft travel from Earth, the tendency of machines to interact in unexpected ways, and the difficulties of repair once missions are under way make cost control difficult as experts prepare missions for flight. From this reality has risen a belief, widely held by people who design and fly spacecraft, that mission success is more important than cost control. This belief was reinforced by the experience of the Cold War, in which big budgets created an assumption that politicians would spend any amount necessary to make the United States "first in space."

The latter point—the belief in a cornucopia of funding—was never true. Belief triumphed over reality, however, and the people who worked on spacecraft created a culture in which cost control was not held in high regard. As a result, cost reduction is not only technically difficult, but generally ignored by people in an aerospace culture where economy and reliability are viewed as polar opposites.

Textbooks in product design commonly instruct students to make trade-offs between various facets of items being designed. Aerospace engineers accomplish this through what they call trade studies. Cost control is traded for reliability. Compressed schedules are purchased with increased spending. The hardware that carried Americans to the moon had to be made lighter to fit the lifting power of the *Saturn V* rocket, and as one textbook generically explained, "this actually will probably increase manufacturing cost."[1]

Since the beginning of the space age, a few people inside the aerospace community have fought the dominant culture. They have sought to reduce the cost of spaceflight by promoting reusability, common components, and design reform. Prior to 1992, all of these efforts failed.

ILLUSION OF CORNUCOPIA Most experts believed that the cost of sending the first Americans to the moon would begin somewhere around $20 billion. About $9 billion would be needed to start building *Saturn V* rockets. About $7 billion would be needed for spacecraft development. Facilities and operational support would consume another $5 billion. Yet when asked for a preliminary cost estimate, NASA Administrator James Webb told a congressional committee that "some people use a number as high as $40 billion."[2] The incident became part of spacefaring lore, a story that served to spread the belief that members of the U.S. Congress would pay any price to put Americans on the moon first. In fact, Congress appropriated the lower sum. Lawmakers repeatedly cut NASA's budget, and White House officials during the administration of Lyndon Johnson very nearly abandoned President John F. Kennedy's end-of-the-decade goal because of cost concerns.

The most serious assault occurred in 1967, when budget director Charles Schultze proposed a $600 million cut in NASA spending. Schultze admitted that his cuts would prevent the United States from achieving Kennedy's deadline, but advised that "it would be better to abandon this goal now in the name of competing national priorities, than it give it up unwillingly a year from now because of technical problems."[3] Budget battles exhausted NASA Administrator James Webb and contributed to his decision to resign 10 months before the triumphant landing on the moon.

Dwight Eisenhower, the first president to preside over the space race, opposed large spending on human spaceflight. Eisenhower viewed proposals for lunar expeditions as a waste of precious tax funds needed for more pressing national security objectives, such as the establishment of a network of military reconnaissance satellites. He dismissed calls for a moon race as a "multi-billion dollar-project of no immediate value."[4]

During the 1960 presidential campaign, Kennedy criticized Eisenhower for creating a "missile gap" with the Soviet Union. No missile gap existed, as Kennedy later learned, but the issue became a focal point for public dissatisfaction with Eisenhower's cautious approach to rocket and space spending. Space advocates hoped that Kennedy and his vice presidential nominee, Lyndon Johnson, would increase spending for rockets and space exploration once elected. Privately, however, Kennedy worried about the possibility that space race spending would "mount radically." He worried during the 1960 presidential campaign, and he worried again after a 1961 speech in which he challenged Americans to race to the moon. Sobered by cost estimates for the lunar expedition, Kennedy secretly approached Soviet premier Nikita Khruschev scarcely one month after his May 25, 1961, speech and asked him to call off the race. Kennedy made the proposal at the June 1961 Vienna summit and publicly repeated the proposal at a September 20, 1963, United Nations speech calling for a joint U.S.-Soviet mission. Ex-president Eisenhower endorsed the proposal. A joint venture, Eisenhower agreed, would "be much cheaper economically."[5]

Members of Congress viewed space spending cautiously. Publicly, lawmakers attacked Eisenhower when he was president for proposing what some called a "beginner" space program.[6] Legislatively, they actually cut his first request for NASA spending. Behind the public facade of support for the moon race, lawmakers cut President Kennedy's space requests in every year—6 percent in fiscal year 1962, 4 percent in fiscal year 1963, and a whopping 11 percent in fiscal year 1964. In the fiscal year 1964 battle, lawmakers on the floor of the U.S. Senate reduced the amount that their own appropriations committee had recommended for NASA spending, a serious vote of no confidence in their budget review process. The final congressional appropriation was $612 million less than President Kennedy had requested and $251 million less the bare minimum that NASA Administrator James Webb said he needed to keep the moon race on schedule.[7]

Like their representatives in Congress, Americans in general approached space spending with chronic ambivalence. They loved the government space program, but were reluctant to pay large sums to carry it out. In a 1961 Gallup public opinion poll, 52 percent of Americans agreed that humans would reach the moon within 10 years, and 58 percent said Congress should not spend the money to go there.[8]

Politicians felt no obligation to support costly spaceflight initiatives beyond the landings on the moon. In 1965, NASA officials unveiled proposals for a post-Apollo space effort. NASA spending that year totaled 4 percent of all federal revenues—a tenfold increase over the sums available for space exploration only five years earlier. Space advocates requested billions more for an Apollo Applications Program that featured a Skylab orbital workshop and a post-Apollo program full of space stations, orbiting telescopes, lunar bases, and expeditions to Mars. Thomas Paine, who succeeded Webb as NASA administrator, predicated his campaign for the new goals on the assumption that federal officials would continue to direct 4 percent of total revenues to space exploration.

The campaign absolutely failed. President Johnson's budget examiners cut funding for the Apollo Applications Program tenfold. President Richard Nixon's budget examiners told Administrator Paine that he would have to run the U.S. space program with just one-third of the funds he needed to achieve his grand vision of exploration. Paine appealed; the appeal failed. In 1971, White House officials removed Paine as NASA administrator, in part because of his relentless pursuit of higher spending.

REUSABLE SPACECRAFT People close to the space program understood the implications of these budget battles. Unless rocket scientists cut the cost of spaceflight dramatically, exploration dreams would stall.

The most promising approach seemed to be some form of reusable space transportation that could cut access cost to space. Instead of launching rockets that crashed into the ocean after each flight, experts proposed to build spacecraft with wings that could land like aircraft and be reused for future flights.

In 1972, President Nixon approved NASA's request to start development work on a reusable space shuttle—the only major new initiative he allowed NASA to pursue. NASA officials agreed to develop the shuttle at a cost of $8.3 billion. The cost ceiling was divided into two parts: $5.5 billion for phase one and $2.8 billion for phase two.[9] For this expenditure, NASA officials promised to build five space shuttles that could be launched for about $10 million (1971 dollars) each time one flew and carry 65,000 pounds to low Earth orbit. This was about one-tenth the cost per pound of launching payloads on expendable launch vehicles like the *Saturn IB*, a worthy objective if it could be achieved.[10]

Unlike Project Apollo, whose funding appeared to be generous, the space shuttle program required NASA officials to develop a new initiative under severe cost goals. A lengthy economic study commissioned by NASA officials and completed by Mathematica, Inc., concluded that only this particular combination of development and launch costs on a reusable launch vehicle would prove cost-effective.[11] More important, the numbers were small enough to win the consent of White House officials. Space shuttle spending rates fit within the annual budget ceiling that White House budget examiners had imposed on civil space employees.

Contrary to the popular perception of massive cost overruns on the early development effort, NASA engineers completed phase one at close to the original goal. Actual expenditures for phase one totaled $5.9 billion in inflation-equalized 1971 dollars, reasonably close to the original goal. (In 1972, White House officials reset the phase one goal at $5.15 billion.)[12] Phase two, however, was a different story. NASA officials had agreed to complete the phase two development program for $2.8 billion. By 1990, they had spent a grand total of $18.3 billion on phase two shuttle development. This sum did not include additional funds provided by the Department of Defense for their share of shuttle development. Even adjusting for inflation ($6.7 billion in 1971 dollars), this was considerably more than allocated. Development work on this scale was never contemplated in the original plan.

NASA officials planned to spend about $10 million on flight costs every time they flew a shuttle mission. The $10 million goal, reset in the value of 2000 dollars, equaled $48 million per launch. Operational cost estimates grew from $10 million (1971 dollars) to $23 million (1975 dollars) to $38 million (1975 dollars) to $57 million (1975 dollars) as experience accumulated. By 1985, the estimated cost per flight had grown to $121 million (current year dollars). That increased to $225 million in 1990 and $393 million by 1994 and $453 million by 2000 (current year dollars). Just to fly, the shuttle cost nine times as much as the original cost estimate, adjusted for inflation. The "factor of 10" cost savings goal evaporated.[13]

The full expense of a single shuttle flight consists of development outlays amortized over the useful life of the vehicle added to operational costs. Analysts who have performed these calculations set the total cost per flight of each shuttle mission at nearly $1 billion.[14] NASA officials used the argument of cost-effectiveness to win approval for a reusable space shuttle. Once they started work on the new spacecraft, cost constraints tended to disappear.

The actual shuttle, as astronaut Michael Collins has observed, is a "tender technology." Its operational economics hardly conforms to that of a commercial airline company, the analogy often employed by its advocates.[15] The shuttle requires thousands of people to prepare it for launch. By 1997, NASA officials had assembled over 20,000 people to prepare each shuttle launch: 18,000 contractors and 2,400 civil servants. Technicians on multiple shifts worked 127 days rebuilding engines, inspecting tiles, installing payloads, and making repairs.[16]

Some spaceflight advocates blame high costs on the infrequency of flights. If only the government would launch the shuttle more often, it would pay for itself, some say. The original economic model prepared by Mathematica assumed that NASA would launch a shuttle every one to two weeks-a total annual rate of between 25 and 50 flights per year. It is true that the marginal cost of flying the shuttle drops quickly as the number of flights increases. NASA flew the space shuttle four times in fiscal 2000. The marginal cost of flying one additional mission that year was about $70 million. Flying more often reduces per mission costs, but not to the "factor of 10" levels anticipated by proponents of the system. Launch costs are still too great to meet that goal even if the shuttle would fly frequently.

Early shuttle advocates wanted to adopt a different technology with different cost parameters. The most popular alternative was a fully reusable shuttle. People promoting the fully reusable concept envisioned two winged rocketships, each flown by a human crew. Pilots would fly the first stage with an orbiter attached to an altitude of 40 miles, then release the orbiter to complete the journey into space. Both craft would land on conventional runways like normal aircraft.[17]

The concept was technologically breathtaking and incredibly expensive. Cost estimates from the Marshall Space Flight Center set development costs for the two-stage, fully reusable shuttle at more than $12 billion, which was certainly understated.[18] Overall operational costs would have increased as well, which is why Mathematica, Inc. economists and White House budgeters concluded that the fully reusable alternative would never work.

Given the technologies at hand, aeronautical engineers could not build a large reusable rocketship with a 65,000 pound payload capacity and fly it for one-tenth the cost of an expendable launch vehicle. NASA executives could have told White House officials and lawmakers in Congress that a big, low-cost space shuttle was not feasible at that time. In the minds of spaceflight advocates, this admission carried a significant risk. They worried, rightly so, that cost-conscious politicians might shut down the human spaceflight

program if such an admission appeared. Rather than risk this un-
thinkable consequence, flight engineers followed a practice widely
accepted at the time. They built a reusable vehicle that worked (it
flew like a rocket and landed like an airplane) in the hope that people
would forget about the original cost goals.

The old cultural assumptions were at work. The people who built
the shuttle paid for higher performance with added cost. They also
stretched out the development and launch schedule. Failure to meet
cost goals, spaceflight engineers assumed, would be forgiven once
the shuttle flew.

COMMON COMPONENTS One of the people who worked hard to get
the shuttle approved was NASA Deputy Administrator George M.
Low. As a career government employee, Low was the highest-
ranking civil servant in the civil space agency. (The NASA administrator is a political appointee.) Low was born in Austria, from
which his family fled after the Nazi takeover. Educated in the United
States as an engineer, he joined the National Advisory Committee
for Aeronautics (NACA) in 1950 and moved through assignments
of ever-increasing responsibility. After the *Apollo 204* fire that killed
three astronauts in the winter of 1967, NASA executives appointed
Low to take charge of the floundering Apollo spacecraft program.
In December 1969, he became NASA deputy administrator, the
agency's general manager.

Shaken by the budget battles over the space shuttle and the post-
Apollo space program, Low told his colleagues that they had to
reduce the cost of spaceflight. "The high cost of doing business in
space . . . places severe restrictions on the amount of productive work
that NASA can do. . . . It therefore becomes an item of first order
business for each of us to find ways to drastically reduce the costs of
all elements of space missions."[19]

Low wanted to accomplish this through the development of
standardized parts, "each with a goal of low cost and high reliability."[20] Up until that time, NASA and its industrial partners had

produced customized components individually tailored to fit the requirements of each spacecraft they built. With each new spacecraft, engineers designed and tested new components. Not only was this costly, but it compromised reliability because experience did not accumulate on items that were used only once or twice.

Low pointed to spacecraft tape recorders as a prime example of this process. Tape recorders, according to a special task force, were "the most failure-prone component in U.S. spacecraft."[21] Industry workers produced different tape recorders for different projects. The small quantity of recorders produced for each project cut into the time available for testing, which in turn reduced reliability.

If he could convince project managers to use common components, Low foresaw, he could break NASA's tradition of exchanging money for reliability. Production lines would develop for low-cost, rugged components. The components might be heavier than they needed to be, but Low reasoned that this problem would be resolved by the much-anticipated falling cost of space access. To further his vision, Low established a special Low Cost Systems Office at NASA headquarters in 1973.[22]

The Low Cost Systems Office conducted studies and held workshops on cost control. Office employees encouraged engineers to develop a number of common components. Engineers at the Goddard Space Flight Center developed a Multi-Mission Modular Spacecraft, which contained a standardized spacecraft bus, standardized transponder, and standardized inertial reference unit. On the whole, however, the impact of the office on civil space spending was minimal. One study set overall savings at just $18.8 million per year.[23] George Low retired from NASA in 1976; the Low Cost Systems Office was abolished in 1979.

Why did the effort falter? People intimate with spacecraft development point to a logical inconsistency in the premise underlying the initiative. Common components reduce the overall expense of producing that particular component, but drive up the costs of adapting the missions that use it. For items like tape recorders, commonality savings outweigh adaptation expense. But for items like rockets designed to boost payloads from low Earth orbit to higher altitudes, the reverse is true. As one aerospace executive observed, "the cost of adapting to many missions can easily overwhelm the savings of a common design."[24]

The effort also failed because managers at the NASA field centers ignored it. The real power to alter attitudes about cost rests with managers directing specific projects large and small. Employees at the Low Cost Systems Office in Washington, D.C., controlled no projects. They were a staff organization at NASA headquarters with no project responsibility and, as such, a weak source from which to alter NASA's overall culture.

DEFINITION BEFORE DESIGN While the effort to cut the cost of spaceflight through reusability was still under way, NASA officials undertook yet another cost reduction initiative. This time they attached the initiative to a specific project, the International Space Station, announced by President Ronald Reagan in 1984. NASA executives set the cost of the space station effort at $8 billion (1984 dollars). This was considerably less than the figure issued by agency cost estimators. A special Concept Development Group, headed by Marshall Space Flight Center engineer Luther Powell, set the cost of the space station closer to $12 billion (1984 dollars).

NASA executives hoped to hold space station expenditures at the $8 billion level through a number of techniques. Most important, they hoped to constrain costs by defining requirements before designing configurations. The inspiration for this reform came from a 1981 report on cost growth conducted by Langley Research Center Director Donald P. Hearth. Hearth's report dealt with the tendency of NASA projects to exceed cost goals established in original planning estimates. His group studied 13 projects that had experienced cost growth varying from 1 to 127 percent.

One of the most significant contributors to cost and schedule growth, Hearth concluded, was "inadequate definition of the technical, cost, and schedule requirements for the project" prior to the decision by officials in NASA, the White House, and Congress to proceed with development. "Current NASA policy," he noted, "does not require pre-project analysis and definition."[25]

John Hodge, director of the Space Station Task Force that prepared the case for project approval, told members of his group to concentrate on requirements. "I wouldn't let anybody draw any pictures," Hodge recalled. "People got very mad at me, because of course there were a lot of engineers on the Task Force, mostly engineers, and they want to start designing things and drawing pictures." Hodge's approach was endorsed by Philip Culbertson, special assistant to the NASA administrator overseeing the space station planning effort.[26]

NASA planners set aside an initial appropriation of $350 million to be spent during fiscal years 1985 and 1986 for what they called the definition phase. The money would be spent analyzing missions the space station would perform, deriving requirements, and formulating a preliminary design. Only after this was done would detailed design work and fabrication begin. In the minds of cost-conscious executives, a very small amount of spending in the beginning ($350 million) would maintain the $8 billion ceiling in the long run.[27]

Project officials spent $346 billion on space station definition, but not with the expected result. The actual definition process did not take place at NASA headquarters, where Hodge and Culbertson worked. Instead, it took place at the Johnson Space Center under the direction of a special engineering team made up of representatives of the old NASA culture. They designed a space station that cost 80 percent more than the original $8 billion estimate.[28]

Very little of the 80 percent increase went for enhanced space station capability. Most of it was directed away from space station hardware to what are known as nonprime activities and cost reserves. Hodge and Culbertson fought hard to restrain such spending. Project managers in the field fought back. Project managers use cost reserves as a hedge against technological uncertainty, thereby helping to improve reliability. Nonprime spending enhances what center directors call institutional capability. Field center executives like to charge programs an "institutional tax" which they use to improve center facilities. Industrial contractors consume nonprime funds by purchasing support equipment and paying for systems engineering, sometimes called contractor "wraps." Eighty-six percent of the increase in the space station budget consisted of nonprime activities and cost reserves.

Engineers from the NASA field centers challenged the commitment of NASA executives to a low-cost space station. The old NASA culture of monetary largess reasserted itself. Leaders of the space station initiative could not fight back because they had disappeared. NASA Administrator James Beggs was forced to resign, the result of a Justice Department investigation into his previous work with the General Dynamics Corporation (the charge later turned out to be groundless). Culbertson left his position as director of the Space Station Program Office and became NASA general manager, a post he relinquished when the new NASA administrator arrived. People in the field centers forced Hodge to retire.

During Project Apollo, a strong program office at NASA headquarters under General Samuel C. Phillips counterbalanced technical challenges from the field. Project Apollo worked because of the tension created between the strong central office and technically competent field engineers. In the space station project, the strong central office collapsed due to pressures from the field.

Space station costs, as a consequence, spun out of control. Government and industry employees spent the entire $8 billion allocated to build the facility without producing any appreciable station hardware. They spent the money on space redesign studies. The space station program endured four major redesign efforts between 1987 and 1993. Congressional committee members and their staff intervened in the design effort, creating a cycle of micro-management that drove up costs and created schedule delays. By 1997, funds devoted to prime activities had dropped to less than 40 percent of total program expense. In 1993, government officials reset the size of the U.S. contribution to the International Space Station at $17.4 billion, an estimate that grew to $26.1 billion seven years later. This was stated in terms of the real amount of money needed to complete the facility, not the long since disappearing value of 1984 dollars. The $26.1 billion did not include the $10.2 billion spent between fiscal years 1985 and 1993.

In the beginning, people running the space station program did not include the expense of transporting components to orbit. They believed, at the time, that the cost of flying the space shuttle would be so low as to be negligible. The space station will require about three dozen flights of the space shuttle during the assembly process. Those funds were not included as part of any of the estimates cited above.

In a modern application of Murphy's law, almost everything that could have gone wrong with the effort to produce a low-cost space station did. Forty years after its creation, NASA officials found themselves still without a proven method for reducing the cost of spaceflight. They had many good ideas, but few that seemed to work.

THE PHILOSOPHY

The White House space policy staff created the "faster, better, cheaper" concept as part of its effort to reduce the cost of President Ronald Reagan's Strategic Defense Initiative (SDI), a Defense Department undertaking popularly known as Star Wars. The concept migrated to NASA after President George Bush in 1989 created the Space Exploration Initiative (SEI), which aimed to establish a lunar base and an expedition to Mars. White House officials viewed NASA cost estimates for SEI as excessively high.

The "faster, better, cheaper" concept was championed by NASA Administrator Daniel Goldin, appointed by the White House to that post in 1992. White House officials selected Goldin because as an aerospace executive he championed smaller, low-cost spacecraft. Officials in the White House wanted Goldin to reform NASA and deconstruct the old culture of high spending. Initially, Goldin sought to do this by building up the agency's small satellite effort and by promoting a new initiative called the Discovery program.

WHITE HOUSE INITIATIVES In the spring of 1983, President Ronald Reagan proposed that the United States develop a space-based defense system that could shield the nation from a nuclear missile attack. From the beginning, political support for SDI was thin, even in the Defense Department, where SDI directors were isolated from regular line activities. Champions of SDI, or Stars Wars as the media dubbed it, needed to prove the feasibility of the concept as quickly as possible to counteract mounting skepticism. To do this, they decided to conduct an experiment in which they would intercept the second stage of a *Delta* launch vehicle during powered flight. Early cost estimates for this experiment ranged from $300 million to $1 billion, with three to five years required to prepare it.[1] This was exactly the sort of cost and schedule largess that critics happily used to attack the Star Wars concept. Directors of the small SDI organization challenged scientists at the Johns Hopkins University Applied Physics Laboratory to develop an intercept experiment in just 14 months, at a cost of $150 million (1986 dollars). The experiment, known as *Delta 180*, was carried out in September 1986, 18 months after approval of its design. *Delta 181* and *183* followed.

Six years after President Reagan had proposed SDI, newly elected President George Bush proposed that NASA return to the moon and organize a human expedition to Mars. Bush reestablished a National Space Council within the White House to oversee both Star Wars and what became known as the Space Exploration Initiative (SEI).

As initially occurred with SDI, SEI excited the supporting bureaucracy to propose grandiose and costly schemes. Although an official cost estimate was never issued, reports of a $400 billion price tag circulated through the space policy community.[2] From the point of view of officials on the National Space Council, NASA civil servants used SEI to demand additional funds for existing programs like the International Space Station. Officials on the National Space Council asked NASA civil servants to prepare a more innovative plan. Civil servants responded with a 1989 planning document known as the 90-Day Study. (The study took 90 days to prepare; the missions it envisioned took up to 20 years.) The 90-Day Study angered staff members on the National Space Council, who found its proposals costly and traditional. According to Stephanie Roy, "the 90-Day Study convinced the Space Council that some sort of cultural change would have to take place at NASA if SEI were to succeed."[3]

In mid-1990, the staff director for the National Space Council, Mark Albrecht, published an article calling for a revival of the U.S.

space program. "The basic goal," Albrecht wrote, "is to do things faster, cheaper, safer, better."[4] Albrecht deliberately inserted the word "better" so as to emphasize that he expected NASA officials to develop innovative approaches to their exploration activities, not just cut cost. The article appeared as part of a symposium on national space policy in *Roll Call*, the newspaper of Capitol Hill.

National Space Council staff members pointed to SDI as an example of the approach they wanted NASA to follow. As if to show NASA officials how this was done, leaders of the SDI organization proposed to develop a spacecraft that would fly to the moon and the near-Earth asteroid Geographos. They called the project Clementine. Normally, a civil mission such as this would be developed by people in NASA, the civil space agency. SDI officials justified their entry into civil space exploration by arguing that the probe would serve as a proving ground for SDI, notably the advanced technology effort underpinning Brilliant Pebbles.

Advocates of Brilliant Pebbles envisioned a large array of satellites capable of detecting and destroying intercontinental ballistic missiles launched toward the United States. To deploy such a large array, project managers had to develop "eyes" that were small and inexpensive to produce—lightweight imaging sensors capable of detecting fast-approaching targets with bulletproof precision. Advocates for the Clementine project proposed to use the moon, the asteroid, and the spacecraft's interstage adapter as targets to demonstrate the performance of their lightweight sensors.

Staff members on the National Space Council also sought to change NASA's culture by changing its leadership. In 1991, Michael Griffin moved out of the SDI organization, where he worked as the director of technology, and moved into NASA, where he took charge of the Space Exploration Initiative. Colonel Simon "Pete" Worden left the Space Council staff to take Griffin's place at SDI, tightening the rope binding together the so-called star warriors on the Space Council and the SEI/SDI staffs.[5]

The ultimate altercation occurred in early 1992, when members of the Space Council engineered the removal of Admiral Richard Truly as NASA administrator. Truly, an ex-astronaut, was viewed by members of the Space Council staff as too committed to NASA's tradition of large and costly activities. Members of the Space Council staff wanted to replace Truly with Lieutenant General James Abrahamson, the first director of SDI. This was too much for congressional Democrats who opposed Star Wars in general and feared the influence of "star warriors" like Griffin and Abrahamson in particular. Members of the Space Council looked for someone else who shared their philosophy, and discovered Dan Goldin. As an executive at the TRW Corporation, Goldin had established a reputation as an advocate of small satellites and had worked on the Brilliant Pebbles initiative, both of which had brought him to the attention of the National Space Council staff.

Work on the Clementine mission continued. The spacecraft, which weighed about 1,000 pounds fully fueled, was launched on January 25, 1994. It completed a detailed mapping of the moon and recorded a radar signature at a large crater at the lunar south pole consistent with the presence of water ice. After leaving lunar orbit, during flight maneuvers designed to send the spacecraft to the asteroid Geographos, a thruster malfunctioned and the probe spun out of control. Like the miner's daughter after whom the *Clementine* spacecraft was named, she was "lost and gone forever."

Experts traced the errant spin to a software error. In spite of the failure of the mission's second phase, the *Clementine* spacecraft was widely hailed as a successful demonstration of the "faster, better, cheaper" approach. The mission had cost only $80 million, including launch and operations, and had made its way from design work to lift-off in just 22 months.[6] *Clementine* marked the first return of an American space probe to the moon in 22 years, and it located a potential resource that could be converted into rocket fuel and other consumables by anyone wishing to venture to that nearby orb.

GOLDIN'S REFORM Cultural changes in organizations do not occur unless they are led by the people at the top. It is nearly impossible to create cultural change in an existing organization by pushing from the bottom up.

The principal spokesperson for cultural change within NASA was Administrator Daniel S. Goldin. Goldin was confirmed in the spring of 1992, in the last full year of the Bush administration. He was not the first choice of the people on the National Space Council screening the nominees, but he was the most acceptable. "Over thirty years ago," he told NASA employees on his confirmation, "I sat down with my father and we filled out an application for the Lewis Research Center that started my career in civil space." Never did he think that someday the president of the United States would fulfill what he called a "boyhood dream" and ask him to lead the civil space agency.[7]

Goldin spent five years at the Lewis Research Center during the height of the Apollo era, when NASA was at the peak of its power and capability. He left the Lewis Center in 1967 for a career with the TRW Corporation, where he specialized in the development of robotic spacecraft. He helped build a variety of scientific, communication, and reconnaissance satellites, and worked on the Brilliant Pebbles initiative.

Goldin's enthusiasm for small, inexpensive spacecraft irritated people within the national space community. He was upset with what he saw as NASA's transformation into a bureaucracy resisting necessary change. "Spacecraft under development today weigh tens of thousands of pounds," he told an aerospace seminar four years before his nomination, "and will cost billions of dollars if something is not done." Unless the government changed its way of doing business, he said, costs would "grow to the point where critical systems are unaffordable."[8]

In spite of his prior attacks on large, expensive spacecraft, Goldin did not approach his confirmation hearings with a commitment to cost-cutting as his chief administrative goal. He was initially cau-

tious about the applicability of the "faster, better, cheaper" philosophy that members of the National Space Council had been promoting. When asked by Tennessee Senator Albert Gore to comment
on proposals for space missions that were "smaller, quicker, and
cheaper," Goldin replied cautiously. "You cannot do every mission
with a smaller satellite," he said. "If we say we will use small satellites to do the total NASA robotic mission, we would be making
a very grave mistake." Upon becoming NASA administrator, he
launched a large series of management reforms, promoting initiatives like Total Quality Management and performance measurement.[9] He did not mention "faster, better, cheaper" in his first address to NASA employees.

All that changed with Goldin's exposure to the NASA budget.
Shortly after arriving on the job, Goldin had to prepare the agency's
budget requests for fiscal year 1994. Dealing with the numbers in
detail convinced Goldin that he had to radically change the way in
which the agency financed its activities: "When I arrived at NASA
seven weeks ago, the agency had set up budget expectations that were
unrealistic and that could not be supported by the President and
Congress in these tough fiscal times. Yes, we might have been able to
get through FY 1993—if Congress provided the President's budget
request—but the budget requirements for the outyears were out of
sight."[10]

Goldin assembled a set of management teams and told them to
rethink the content, scheduling, and technology of NASA programs. And he began to talk about a new initiative. In remarks to
workers at the Jet Propulsion Laboratory, he told them that the
management teams would tell employees "how we can do everything
better, faster, cheaper."[11] Beginning with the planetary program, he
said, NASA had to stop building large spacecraft.

There's a paradox at work here that creates a downward spiral.
Launching fewer spacecraft means scientists want to pile every instrument they can onto whatever's going to fly. That increases the

weight, which increases the cost of the spacecraft and the launcher. Fewer spacecraft also means we can't take any risk with the ones we launch, so we have to have redundancy, which increases weight and cost, and we can't risk flying new technology, so we don't end up producing cutting edge technology.

A philosophy like that, Goldin complained, would produce a program "that takes 10 years to get . . . approved, 10 years to build and launch, and up to 10 years to reach its destination." It was no wonder, he concluded, that the space agency had lost support among graduate students, scientists, Congress, and the American people. Occasionally, there might be a scientific reason to build larger spacecraft because of the need for aperture size, stopping power, or simultaneity of measurement, "but this would be the exception, not the rule."[12]

Goldin's ideas about "faster, better, cheaper" evolved quickly as he gained experience as NASA administrator. In speeches and interviews he laid out concepts to guide implementation of the initiative. "We have begun a cultural revolution," he professed, "that will take years to fully complete but will provide a wealth of benefits."[13] Borrowing the title from a popular management book at the time, *Reinventing Government*, Golden announced that "we are trying to see if we can re-invent NASA."[14]

To begin the revolution, he said, NASA and its contractors had to start with a shared vision, including both the ultimate purpose of their enterprise and the methods for accomplishing it. "We should send a series of small and medium sized robotic spacecraft to all the planets and major moons, as well as some asteroids and comets," Goldin announced, identifying where the initiative would start. The spacecraft should be smaller, cheaper, and faster. "Let's see how many we can build that weigh hundreds not thousands of pounds; that use cutting edge technology, not 10-year old technology that plays it safe; that cost tens and hundreds of millions, not billions; and take months and years, not decades, to build and arrive at their destination."[15]

Increase the frequency of launches, he said. "Let's start launching probes every year—more than one a year," he told Jet Propulsion Laboratory employees. "We can't let grad students turn into senior citizens as they wait for their work of a lifetime to get launched, or we'll end up with no one pursuing space as a career."[16]

Reduce spacecraft cost. "Slice through the Gordian knot of big, expensive spacecraft that take forever to finish," he urged. Cut development costs by building smaller spacecraft. Cut launch costs by placing smaller spacecraft on smaller rockets. Cut operational costs by flying smaller spacecraft to their destinations faster. Goldin enouraged engineers to put instruments on a common spacecraft chassis. "By building them assembly line style, we can launch lots of them, so if we lose a few due to the riskier nature of high technology, it won't be the scientific disaster or blow to national prestige that it is when you pile everything on one probe and launch it every ten years."[17]

Smaller, less expensive spacecraft do not have to be inferior to larger ones. Goldin fought the mind-set that equated cheaper with inferior. Cheaper could be better. He railed at engineers who thought they had to cut capability or testing in order to cut the cost of a project. "Costs are a paradox," he insisted.[18] Clever people could reduce cost and get a better result if they were willing to push the limits of technology and take more risks.

Goldin told project managers to take more risks. "A project that's 20 for 20 isn't successful," he said. "It's proof that we're playing it too safe." Developers of automated spacecraft could afford to lose a few. "You don't have the same safety requirements that the human space program does. That means you should push the limits of technology even more." If the gain was great, risk was warranted. "Let's be bold and not be afraid." he said. "It's OK to take risk when you're pushing the frontiers of the possible." Frequent launches would take the sting out of failure. "With a launch a month, you could go to the cutting edge and you could lose three or four spacecraft a year. It would be okay because you keep launching."[19]

Push the limits of technology. "There is this mind-set that says if you want to be low cost, use commercial off-the-shelf technology and don't try to invent. I don't necessarily agree with that."[20] As a young NASA engineer, 30 years earlier, Goldin had worked on the problem of electric propulsion. NASA had still not flown an electric propulsion spacecraft. "We've got technology sitting in your labs that people drool over elsewhere," he told NASA employees. Use it, he implored. Incorporate it into spacecraft. Transfer it to industry so that the United States becomes more competitive worldwide.[21]

THE FOUNDATION When Daniel Goldin arrived at NASA, he found only a few people working on low-cost endeavors. A group of space scientists had attempted to revive the agency's small satellite program. Another set of officials in NASA's space science division had put together a special initiative designed to increase flight frequencies, utilize new technologies, and reduce costs. They called it the Discovery program. Goldin examined the fledgling program and called it "the world's best kept secret."[22]

NASA employees had been launching small satellites since before they worked for NASA. *Explorer 1*, the first U.S. satellite, was an 18 pound satellite developed by the Jet Propulsion Laboratory and launched on a *Juno-1* rocket built by the Army Ballistic Missile Agency, both organizations destined to become part of the new civil space agency. The satellite was so small that three mission managers could raise a full-scale model above their heads for a famous photograph celebrating its launch. The satellite carried a cosmic ray detector and two small radios to transmit readings back to the Earth. It made a most important discovery, detecting a belt of high radiation at an altitude where rocket designer Wernher von Braun wanted to place an Earth-orbiting space station. The radiation belts were named after James A. Van Allen, the physicist from the University of Iowa who built the instrument carried onboard.

Explorer 1 had to be developed in an incredibly short period of time. Prior to its launch, officials in the Eisenhower administration had assigned the U.S. satellite effort to another organization, the Naval Research Laboratory. Scientists there had been preparing a *Vanguard* rocket and a 22 pound satellite for more than two years. On December 6, 1957, the *Vanguard* rocket rose 3 feet off its launch pad at Cape Canaveral, shook briefly, and disappeared in a spectacular ball of flames. Anxious to match the launch of the first Soviet Earth satellite, Eisenhower had allowed the von Braun-JPL-Van Allen team to proceed with *Explorer 1* as an alternative should *Vanguard* fail. The von Braun-JPL-Van Allen team set up the mission in just 84 days and sent the satellite into orbit on January 31, 1958.[23]

The success of *Explorer 1* launched a decade of small satellite activity. The United States launched 41 *Explorer*-class satellites between 1959 and the end of 1969. Project Vanguard finally got off the ground on March 17, 1958. Nine additional small spacecraft in the Pioneer series sped into the cosmos during the 1960s. *Pioneer* spacecraft weighed between 13 and 144 pounds. The peak year for small satellites was 1966, when all U.S. organizations combined dispatched more than 60. A small satellite or space probe, by the conventional definition, weighs less than 400 kilograms (880 pounds), including propellants.[24]

The number of small satellite launches in the United States declined sharply as the 1960s wound down. In part this was due to the development of rockets that could carry larger loads. At the beginning of the space age, most U.S. satellites were small because rockets were small. Early versions of the *Thor-Delta* launch vehicle, used to boost many *Explorer* satellites into orbit, were capable of lifting no more than 1,300 pounds into low Earth orbit. With the advent of the space shuttle, rated to carry 65,000 pound loads, the number of small U.S. satellite launches fell to zero. In the Soviet Union, where engineers never developed an operational space shuttle, the frequency of small satellite launches did not decline.

U.S. space experts affected a philosophic shift toward large, multipurpose platforms. This is well represented by the history of NASA's "Mission to Planet Earth," inaugurated in the mid-1980s as a means to track environmental changes and monitor the interaction of land, sea, and air. Space scientists hoped to use *Titan 4* rockets to orbit a flotilla of large automated platforms jammed with scientific instruments into orbits. Each platform would weigh 15 tons and cost nearly $400 million to develop.[25]

Large, multipurpose platforms contain a perverse incentive. Because they are large, they tend to reduce the frequency with which follow-on projects are approved. This creates a situation in which scientists believe that they must place their experiments on projects already under way or lose the chance to fly for long periods of time. The rush to place instruments on existing projects drives development costs beyond initial estimates. Increased costs prompt politicians to stretch out development programs, which further delays new starts. Stephanie Roy, a historian of the "faster, better, cheaper" concept, notes the effect that this has on the space enterprise.

> The perception that new starts were to be few and far between created an environment in which scientists felt compelled to get on board (and increase the science return on) those missions that did get approved. While this is a natural response, its actual result is the creation of a feedback loop in which efforts to get on the last train leaving the station contributed to cost growth in major programs and the crowding out effect on new starts.[26]

The desire to build large, multipurpose platforms began as an effort to save money, based on the philosophy that a shared platform with many instruments would reduce basic spacecraft costs for all. In practice, however, the effort led to very large projects with rapidly escalating costs. Norman Augustine, an aerospace executive and lead author of a 1990 report on the future of the U.S. space program, captured the perversity of this situation in a book entitled *Augustine's Laws*. Augustine directed his criticism at the soaring cost of military aircraft. "In the year 2054," he postulated, "the entire defense budget will purchase just one aircraft." Translating his observation to the space program, aerospace experts predicted that prevailing trends would result in a situation where engineers would be able to produce only a few new spacecraft each year. Augustine meant his observation to be humorous, but to many space exploration advocates the trend seemed grimly real.[27]

Cost escalation and tight budgets encouraged resurrection of the small satellite business. Scientists in charge of the Explorer program, which had evolved toward larger satellites as space exploration matured, renewed their interest in smaller spacecraft. In 1988 they inaugurated the Small Explorer initiative. Program managers conducted their first launch in 1992, a 350 pound satellite sent into Earth orbit on a *Scout* rocket to investigate the composition of local interstellar and solar material. Four more launches occurred in the next seven years, with satellites ranging in size from 422 to 622 pounds. Officials working on Mission to Planet Earth abandoned their large, multipurpose platforms in favor of smaller satellites. The return to smaller missions in the civil space program begun in the late 1980s provided part of the foundation for the "faster, better, cheaper" initiative.[28]

In April 1992, as Dan Goldin assumed leadership of NASA, the Senate Appropriations Committee directed NASA to prepare a plan "to stimulate and develop small planetary or other space science projects, emphasizing those which could be accomplished by the academic or research communities."[29] NASA officials called it the Discovery program. From mission concepts already under study, they selected two projects. Work on the two projects—Near Earth Asteroid Rendezvous (NEAR) and Mars Pathfinder—began in 1993. The officials set up a competitive, peer review process for choosing more projects. The third mission—the first to be selected as a result of open competition—was established in 1995. It was called Lunar Prospector, a follow-on to the Clementine mission designed to create a more detailed map of the moon and look closely for evidence of water ice.[30]

To qualify for the Discovery program, projects had to meet a strict set of guidelines. Total development time could not exceed 36 months. Total development cost could not exceed $150 million and operations costs could not exceed $35 million (1992 dollars). Spacecraft had to be small enough to be launched on a rocket no larger than a *Delta II*, which in 1999 cost about $55 million per launch. Anyone could form a mission team and submit a funding proposal. Mission teams could consist of representatives from universities, industrial firms with experience building flight hardware, and supporting government centers. When NASA officials issued the first call in 1994, 28 proposals arrived at their doors.[31]

NASA officials selected and funded six Discovery missions. By the end of the millennium, four had flown: Near Earth Asteroid Rendezvous, Mars Pathfinder, Lunar Prospector, and Stardust (a mission to fly into a comet's tail and return its dust to the Earth). Mars Pathfinder received more public attention than any other mission that decade. Lunar Prospector detected more evidence of water ice on the moon. NEAR project managers got a scare when they temporarily lost contact with their spacecraft as it approached the asteroid Eros. For a while, it appeared that NEAR would become one of those lost missions that NASA Administrator Daniel Goldin had urged spacecraft managers to risk in their quest for innovation. Said Tom Coughlin, project manager at the NEAR mission control center in Baltimore, "it was the longest 27 hours of my life."[32]

The reception given the Discovery program encouraged imitation. NASA officials applied its guiding principles to the exploration of Mars. Project engineers launched Mars *Global Surveyor* in 1996, then Mars *Climate Orbiter* and Mars *Polar Lander* two years later. As in the Discovery missions, NASA executives capped mission costs and insisted that spacecraft be small enough to fit on *Delta II* launch vehicles. Mars *Global Surveyor* collected detailed data on the planet's geology and confirmed that the planet was once much warmer and wetter. The other two spacecraft disappeared.

NASA executives set up the New Millennium program in 1994. The program was designed to "identify and flight-test new technologies that will enable science missions of the early 21st century."[33] Science fiction became science fact when NASA launched *Deep Space 1*, the first New Millennium project, in 1998. *Deep Space 1* employed an ion propulsion engine, an exotic technology that had formed the basis for a *Star Trek* episode but had never been tested in space. The engine unexpectedly shut itself off a few times when it first started, but ran smoothly thereafter.

Deep Space 2, the second New Millennium project, hitched a ride on the *Polar Lander* to Mars. This amazing project consisted of two baseball-size probes designed to separate from the lander and crash into the Martian surface with such force that they would penetrate to a depth of 2 feet, where they would search for water and communicate back to Earth.

That made nine spacecraft launched by the end of the decade under the "faster, better, cheaper" philosophy. Added to that group were five satellites in the Small Explorer program and two from the Small Satellite Technology Initiative, for a total of 16. The five Small Explorer satellites were designed to produce "extraordinary performance while fully embracing the essence of 'smaller, faster, cheaper,'" a NASA publication read. The technology initiative was designed to make spacecraft less expensive and funnel new ideas into the private sector.[34]

From these latter two groups the first failures emerged. The first two projects in the Small Satellite Technology Initiative were called *Lewis* and *Clark*. The *Lewis* satellite, designed to observe the Earth with sophisticated remote-sensing technologies, was launched in August 1997. As part of the overall effort to reduce operational costs, flight engineers gave the satellite the capability to operate without continuous monitoring by people on the ground. No one was sitting at the ground control center on the fourth night of operations, when the satellite began to tumble. With no one to correct the problem, the *Lewis* satellite drained its batteries and went dead. One month later it dropped back into the Earth's atmosphere and burned up.

The *Clark* satellite never got off the ground. Like its sister craft, *Clark* was designed to conduct high-resolution observations of the Earth. By late 1997, it was over budget and nearly two years behind schedule. In February 1998, NASA executives told the industry group developing Clark to stop work on the project.

The fifth project in the Small Explorer program was called WIRE, for Wide Field Infrared Explorer. The satellite contained a sophisticated telescope capable of collecting infrared images of galaxies in the process of evolution. To shield the telescope from its own heat emissions, the instrument was placed inside a container of frozen hydrogen and cooled to minus 430 degrees Fahrenheit. Project engineers, worried that the telescope's protective cover might separate prematurely following the rigors of launch, installed safety devices to prevent such a mishap. The cover separated anyway. Without its protective cover, the frozen hydrogen warmed up and began venting, and the satellite spun out of control. In the time that it took ground operators to stabilize the spacecraft, the hydrogen had escaped, and the telescope could no longer be used.

The loss of WIRE opened the horrible year for proponents of "faster, better, cheaper." In 1999 project managers lost *WIRE*, Mars *Climate Orbiter*, Mars *Polar Lander*, and the twin *Deep Space 2* micro-probes. Writers at *Science* magazine called the early failures "a bitter blow to NASA engineers and scientists who, at the direction of their chief Dan Goldin, have pushed for cheaper and faster ways to get low-budget and high-tech satellites into orbit."[35]

Advocates of the initiative knew that some missions would fail. Out of the first 16 "faster, better, cheaper" projects scheduled for launch during the 1990s, five failed in space—more than had been expected when the initiative began. Moreover, NASA executives had promised they would terminate any project that incurred cost overruns. When the *Clark* satellite broke the 15 percent barrier, NASA executives did not wring their hands and rescue the program with additional funds. They canceled it, just as they had said they would do.[36]

RESISTANCE TO CHANGE Goldin's advocacy of "faster, better, cheaper" encountered substantial resistance within NASA. Not only did Goldin's proposal threaten NASA's fundamental organizational culture, he had personally replaced a popular NASA administrator and ex-astronaut who represented the agency's commitment to large space endeavors. At the beginning of his term, within the aerospace community at large, Goldin was not popular. Many believed that he and his initiatives would disappear after the Democrats won the 1992 presidential election.

"It's unbelievable," Goldin said, recounting the resistance his initiatives encountered. "Go try and make some change at NASA," he complained. Goldin grumbled that the same people who criticized NASA were quick to resist significant reform. "They inject you with a toxin the day you arrive and that toxin flows through your brains and so long as you deal with the special interests, and you keep the status quo, you're allowed to live and there's a little time bomb. They press the time bomb as soon as you get out of line."[37]

Goldin experienced significant difficulties convincing people in the aerospace community that "cheaper" could also be "better." For many years aeronautical engineers have manufactured what they call Class 3 (or Class C) satellites. A Class 1 satellite is a precise machine with lots of redundancy that is designed to succeed. A Class 3 satellite is cheaper. Its budget is small, its mission is short, and "effects of failure are not catastrophic."[38] To compensate for added risk, engineers may manufacture and fly multiple Class 3 satellites for a single mission. Historically, most small satellites have fallen into the Class 3 category. Many aeronautical engineers associated the challenge of lowering the cost of spaceflight with the manufacture of Class 3 satellites. They know how to do that. They have been building Class 3 satellites for years.

Goldin and other "faster, better, cheaper" advocates struggled to convince aerospace engineers that highly visible missions could be conducted at relatively low cost without suffering the failure rates associated with Class 3 satellites. "Cheaper," in other words, could also be "better." To demonstrate this point, and avoid the fate of previous cost-cutting efforts in NASA, Goldin and other advocates knew that they had to prove the workability of their approach on a difficult mission.

MARS *PATHFINDER*

The most challenging test of the "faster, better, cheaper" initiative involved the Mars Pathfinder mission, launched in December 1996. Between 1962 and 1996, the United States and the Soviet Union sent 19 robotic missions to Mars. Only six successfully completed their objectives. The Soviet Union dispatched 10 missions, including five landing craft. Every Soviet lander failed. Mars, it would seem, eats spacecraft.

The most successful mission to Mars occurred in 1976, when the U.S. *Viking* spacecraft arrived. Viking was a classic NASA mission: complex, redundant, and expensive. The Viking mission cost $1.06 billion—the equivalent of $3.9 billion using the cost of aerospace goods at the time of the Pathfinder mission.[1] The Viking program bought two spacecraft that landed on the Martian soil and two spacecraft that studied the planet from orbits around Mars.

Proponents of the "faster, better, cheaper" initiative asked the Pathfinder team to put a lander and a rover on the surface of Mars for one-fourteenth of the inflation-adjusted cost of the 1976 Viking mission. Viewed historically, it was as if the Viking team had been asked to land on Mars during the 1970s for no more than $72 million. The Pathfinder team, moreover, was asked to design, build, and prepare their spacecraft for launch in just 3 years. The Viking team had taken 6.

Within aerospace circles, the cost and schedule goals given to the Pathfinder team were widely thought to be impossible. The goals provided a major test of the "faster, better, cheaper" concept.

THE MISSION At the time of the project's conception, exploration advocates believed that Mars *Pathfinder* would be the first in a series of small, inexpensive monitoring stations that NASA would spread across Mars. Advocates of one leading concept, called Mars Environmental Survey (MESUR), envisioned a network of 16 monitoring stations around the red planet. MESUR was eventually canceled in favor of alternative endeavors, but the underlying philosophy remained. Exploration advocates wanted to fly missions to Mars every 2 years, when the planets aligned. No one wanted to wait 20 years between missions, the length of time since the *Viking* spacecraft had flown.

To fly frequently, exploration advocates had to fly cheaply. For the cost of one Viking-type mission, NASA could send robotic spacecraft to Mars at every conjunction for 20 years and still have money left over. Exploration advocates asked the Pathfinder team to test this concept. They called it "Pathfinder" because the project blazed the trail for a new generation of low-cost spacecraft.[2]

Build a small environmental monitoring station, exploration advocates told scientists and engineers at NASA's Jet Propulsion Laboratory, that could measure atmospheric conditions and take panoramic pictures in subzero temperatures. Land it directly on the surface of Mars, without orbiting or descent engines, using air bags or a crushable material to soften its impact, a technology that had never been employed for planetary landings. Integrate into the lander an automated rover, another new technology. The rover would roam the surface of Mars, determine the composition of rocks, and communicate with the lander, which in turn would communicate with Earth.

NASA's officially stated cost for the Mars Pathfinder mission, upon completion, totaled $265 million. That paid for a $170 million lander, a $25 million rover, a $50 million launch system, plus $14 million for operations (see box 3).[3] Compared to the Viking mission, the Pathfinder team saved $3.6 billion in inflation-adjusted currency.

CHEAP ACCESS TO SPACE For many years, exploration advocates have touted the benefits of cheap access to space. The cost of transporting

2 **COST OF VIKING PROJECT** (real-year dollars, in millions)	
Lander	609.9
Orbiter	217.1
Project management, integration, and mission operations before launch.	48.2
Mission operations after launch	102.5
Launch vehicles.	78.6
Total	1056.2

materials to low Earth orbit in the last decade of the twentieth century exceeded the price of gold. Launch vehicles delivered water, food, equipment, and propellants to low Earth orbit at a cost per pound that varied from $4,500 to $7,000.[4] The price of gold, as of late 1998, was $293 per troy ounce, or $3,500 per pound. Whoever reduced the cost of spaceflight significantly, advocates proclaimed, would open the space frontier.

The cost of moving goods beyond low Earth orbit, to the planets, is considerably more. In the 1990s, the additional rocketry necessary to escape the Earth's gravity easily pushed total launch costs for planetary probes past $20,000 per pound. In terms of gravitational pull, low Earth orbit was about halfway to anyplace else in the solar system. In terms of cost, it was still a long way to go.

Lacking a revolutionary new launch technology, the Pathfinder team cut transit costs using the only method available to them. They made their spacecraft so light that it could be launched on a small, inexpensive rocket. The fully fueled Pathfinder space probe, with its micro-rover and transit container, weighed slightly more than 2,000 pounds.[5] Mated with a Payload Assist Module (PAM-D) to push it from Earth orbit to Mars, the whole package could be launched on a *Delta 2* model 7925 rocket. The *Delta 2* is a small- to intermediate-size liquid-fueled rocket with strap-on solid rocket boosters capable of pushing 11,000 pound payloads to low Earth orbit. Keeping

3	**COST OF MARS *PATHFINDER*** (real-year dollars, in millions)

Lander

Flight systems 135.3

Science & instruments 13.7

Project management. 7.1

Mission engineering and operations 10.0

Other 4.6

     (Subtotal: Lander . . . . 170.7)

Sojourner rover 25.0

Mission operations after launch 13.9

Tracking and data support 0.7

Headquarters support 4.8

Launch vehicle. 50.3

Total 265.4

spacecraft weight low enough to permit a launch on a *Delta 2* was a major mission goal. Total launch cost to the Mars Pathfinder team was $50.3 million.

The Viking mission, by contrast, required a larger launch vehicle. Each *Viking* probe (there were two) weighed 7,759 pounds. That included one lander, one orbiter, necessary propellant, and the lander containment capsule. The Viking team launched their probes on two *Titan 3* rockets with *Centaur* upper stages.[6] Each launch system cost $39 million in the currency of its day. By using a single *Delta 2* rocket, the Mars Pathfinder team saved $240 million relative to the inflation-adjusted cost of the Viking launch system.

The $240 million launch savings was a considerable sum, but only 7 percent of the total $3.6 billion cost reduction that the Pathfinder team achieved. After launch savings, the Pathfinder team still had to find $3.4 billion more in order to meet their overall cost reduction goals. Cheaper access to space, the panacea of spaceflight advocates everywhere, cannot produce savings of that magnitude. For the next generation of launch vehicles, rocket scientists hope to cut the cost of transporting goods to low Earth orbit to about $1,000 per pound. That would be a remarkable achievement. But it would hardly dent the aggregate savings the Pathfinder team had to achieve on their whole mission.

Assume that the price of a *Delta* rocket could be cut to $1,000 per pound—say to $11 million per launch. That would reduce launch costs by an additional $39 million, an important savings but still a long way from the full $3.6 billion savings the Pathfinder team needed to achieve.

This is a surprising finding. Cheap access to space is not the panacea that exploration advocates profess it to be. It is an important step in the business of cost reduction, but just part of a long-term challenge.

RISK-TAKING To further reduce mission costs, the Pathfinder team accepted risks that Viking team members had been unwilling to endure. Viking team members took extensive steps to ensure that at least one lander would operate as planned on the surface of Mars. To do this, they commissioned the construction of three landers: a third to make sure that at least two got into space and a second to make sure that at least one landed safely on Mars. The two landers that flew to Mars were attached to two orbiters, designed to take high-resolution pictures of prospective landing sites while the landers circled Mars. Viking team members designed an elaborate landing system, with computers, parachutes, radar altimeters, and a throttleable rocket engine. They developed a redundant radio communication system, from the lander to the orbiter and back to Earth as well as from the lander directly to Earth, so that any spacecraft that landed safely could assuredly phone home.

Viking team members agonized over the selection of landing sites. The primary purpose of the *Viking* probes was to search for life in the Martian soil. The most interesting sites were ones that showed evidence of water erosion. Any life that remained on the surface of Mars would have to be near a potential water supply. Ancient water channels, however, were rugged and dangerous places to land. Orbiter photographs helped the site selection process considerably, but nobody knew whether a lander sent to a moderately interesting site would slide down a cliff or hit a boulder or disappear in quicksand. "A crashed lander is not very useful even if it did crash in the most interesting part of the planet," one team member observed.

The Mars Pathfinder team did none of these things. Team members launched one lander, with one radio, and picked a dangerous place for it to land. The landing site was situated at the mouth of an ancient outflood channel, littered with boulders that had washed downstream. The team aimed their single spacecraft, with no backups, straight at the landing site. The spacecraft did not bother to orbit Mars before landing. It sped directly into the Martian atmosphere and, after using a heat shield and parachute to slow itself down, crash-landed at 35 miles per hour. The Mars Pathfinder team cushioned the crash with an inflatable Vectran ball.

The landing plan, in the minds of many spacecraft engineers, was idiotic. It could fail in any of a dozen ways. The space probe in its protective container was designed to hit the Martian atmosphere at 17,000 miles per hour. A heat shield and parachute would slow the probe to 140 miles per hour, creating deceleration forces capable of crushing delicate machinery. At an altitude of about 300 meters, an air bag would inflate around the spacecraft. Small rockets would fire for approximately 2.2 seconds, briefly suspending the whole system about 30 meters above the ground. More fireworks would sever the bridle holding the spacecraft and its air bag to the parachute. The remaining thrust of the rockets would push the parachute away so that it would not tangle with the air bag. The air bag would fall to the ground and bounce wildly before coming to a resting position. Designers added devices that caused the air bag to automatically deflate and the metal petals enclosing the lander to open in such a way as to place the lander in an upright position. Still out of touch with its operators, the lander would automatically swivel its high-gain antenna, find the Earth, and contact the flight team in Pasadena, California, through a ground station in Spain. That, at least, was the plan.

Many people told Pathfinder project manager Tony Spear that the convoluted landing system would certainly fail. In practice, the system worked perfectly, except for a minor glitch when a lander petal got tangled in the deflated air bag.[7] The *Pathfinder* landing system cost $27 million, according to official records, or about 20 percent of the total expenditure for flight systems on the lander.

Pathfinder team members also saved money by producing and flying a single spacecraft. This saved the expense of a second launch vehicle and simplified the process for testing and flying the machine. It did not, however, save much money. Only a small proportion of the funds allocated to space missions are spent constructing hardware, and the effect of building one more or one less spacecraft tends to be incremental. When NASA officials during the Viking era decided to cancel assembly of their third lander, they saved only $5 million.[8] The whole Viking lander development effort cost $357 million in the currency of its day, but the Viking team saved only $5 million by reducing the number of landers by one-third. (The full cost savings of building and testing one less lander totaled $20 million, but three-quarters of that had already been spent when the cancellation decision occurred.)

Risk-taking was an important component of the Pathfinder cost reduction strategy. By itself, however, it probably saved no more than 10 percent of the $3.6 billion they needed to find. To achieve the remainder, members of the Pathfinder team had to take other steps as well.

THE 80/50 RULE When engineers and scientists debate spacecraft design, they frequently wrestle with the exponential effects of incremental changes in capability. This debate often appears in the form of an 80/50 rule, as in "we could conduct 80 percent of the science being proposed for this mission with a spacecraft that would cost only 50 percent as much." The exact numbers vary from project to project, but the general principle remains the same. Less capability translates into substantial cost savings.

Without question, members of the Mars Pathfinder team lowered the cost of their undertaking by reducing the amount of science their spacecraft could perform. The most obvious reduction, relative to the Viking mission, was the absence of any orbiting vehicles. The price tag for the whole Viking mission included two orbiters, which cost a total of $217 million or about $800 million in inflation-adjusted dollars. The orbiters carried scientific instruments that took visual images, measured surface and atmospheric temperatures, and mapped the distribution of water vapor.

The *Viking* landers had more capability than the *Pathfinder* spacecraft. The most expensive scientific instruments on the *Viking* landers were those designed to look for life. There were three: a biology instrument, a gas chromatograph mass spectrometer, and an X-ray fluorescence spectrometer. At $220 million (inflation-adjusted dollars), the biology instrument alone cost more than the entire *Pathfinder* lander. The gas chromatograph mass spectrometer cost $150 million in Pathfinder-year dollars.[9] By comparison, the most complicated piece of equipment on the Pathfinder mission was the *Sojourner* rover, which cost only $25 million.

At the beginning of the program, scientists building the Viking biology instrument estimated that it would cost $42 million (inflation-adjusted dollars). This provides a rough point for comparing its capability with that of the *Sojourner* rover, which cost $25 million. Relative to initial cost estimates, the biology instrument had about twice the capability of the rover. The final cost of the Viking biology instrument, however, was much higher. Due to cost overruns, the biology instrument actually cost $220 million (inflation-adjusted)—5 times the original estimate. The added expense was incurred by Viking subcontractors struggling with an unfamiliar technology, not from differences in capability. Advantages in technology allowed Pathfinder and Sojourner managers to develop scientific instruments without the cost overruns Viking team members incurred.

Critics of the Pathfinder approach insisted that much of the cost savings claimed by the Pathfinder team could be ascribed to different capabilities. Said one member of the Viking lander science team, who confessed to being "a bit weary" of the cost comparisons between Viking and Pathfinder:

> If Viking cost 10 times as much (in inflation adjusted dollars), consider how much more it did. It had four spacecraft, two of which orbited Mars and mapped and analyzed its surface in far greater detail than has been done before or since. At the same time, the other two spacecraft landed safely, analyzing on the way in the molecular and isotopic composition of the atmosphere, not just its pressure and temperature. These two put into operation not one but a total of 10 analytical instruments, most far more complex and sophisticated than Pathfinder's, two meteorological "stations" and two seismographs (one of which admittedly failed to work).[10]

The Pathfinder team clearly saved money by reducing capability. The best analysis suggests that this reduction accounted for slightly more than $1 billion in Pathfinder-year dollars—almost 30 percent of the total cost savings. Most of that—$801 million—was due to the absence of orbiters. The remainder was due to differences in scientific instrumentation, particularly the absence of life detection instruments, and the fact that Viking operated over a longer period of time, which increased operational expenses. Capability reduction due to differences in scientific instrumentation saved about $50 million in inflation-adjusted dollars. Briefer operations saved $188 million in inflation-adjusted dollars, the amount budgeted for 4 years of extended Viking operations.

Even if the Pathfinder team had developed an orbiter, incidentally, they would still not have spent as much money as the Viking team. At the same planetary conjunction as the Pathfinder launch, another JPL team dispatched the Mars *Global Surveyor*. This successful mission cost $131 million for spacecraft development, plus another $143 million for launch support and operations. Viking managers spent $801 million in inflation-adjusted dollars to develop their two orbiters. Powerful forces such as technology and changes in project management worked to lower mission costs in the later time frame. The technology used to produce the imaging systems for the *Viking* and *Pathfinder* landers provides a dramatic example of the way in which better technology cuts costs for equal capability.

TECHNOLOGY Both landers, *Pathfinder* and *Viking*, contained cameras, or imaging systems. Both took remarkable pictures of the Martian landscape, some in color and some in stereoscopic vision or 3-D. The Viking team spent $27.3 million to develop the cameras for their landers, approximately $100 million in inflation-adjusted dollars.[11] The Pathfinder team spent just $7.4 million. Even accepting the fact that the Viking team bought cameras for two spacecraft, the difference is astounding. Much of it was due to advances in technology.

The Viking cameras were bulky affairs, 22 inches tall and at their base 10 inches wide. Each lander held two. The cameras worked by scanning a single vertical line 512 pixels tall, one pixel at a time. A nodding mirror in combination with a set of lens concentrated each light beam onto a single array of light-sensitive diodes. The array contained 12 diodes capable of detecting different focal lengths or colors. The array measured light intensity one pixel at a time and amplified each signal. The signal was converted into a digital format and stored for transmission to Earth. Once the camera finished scanning a single line, it rotated slightly and began scanning another. The camera took about 2 minutes to complete a standard size black-and-white picture. A color panorama of the landing site took about 30 minutes to prepare.

When the project began, Viking managers estimated that the imaging system could be designed and built for $6.2 million. The actual cost ($27.3 million) was 4 times that amount. Technical challenges persisted throughout the design and fabrication phase. "The most potentially devastating problems involved the tiny photosensor array," wrote the leader of the lander imaging science team. Every time industry technicians built one of the 1.3 inch wide arrays, it would fail. Project scientists worried that they had designed an instrument no subcontractor could fabricate, given the state of microelectronics manufacturing in the early 1970s. Scientists recommended that project engineers incorporate a simpler design with less capability. "Fortunately," the lead scientist recalled, "our suggestion was shelved." The technically sophisticated design was retained. Although the resulting images were spectacular, Viking project managers paid for this decision with substantial cost overruns.[12]

The Pathfinder camera sat in a small cylinder, about 4 inches in diameter, mounted on top of a 20 inch extendible mast. At the heart of the imaging system was a CCD detector, which scientists characterize as a piece of "electronic film." The detector contained 63,488 pixels, each capable of recording how much light fell on it. A "snapshot" created an image 248 pixels wide by 256 pixels tall. Motors, filters, and lenses allowed the camera to take a variety of pictures of differing composition. For panoramic images, the camera took many small pictures and pasted them together.

The CCD detector was manufactured by the Max Planck Institute for Aeronomy in Germany. The entire imaging system was built by the University of Arizona Lunar and Planetary Laboratory, under the direction of astronomer Peter Smith. By the mid-1990s, CCD detectors were a well-established technology used by astronomical observatories around the Earth and in space. Because the detector technology was well-established, Smith could concentrate project funds on adding more filters and improving the calibration of the camera, which produced better images. The cameras on the *Viking* and Mars *Pathfinder* landers had approximately the same resolution, but the *Pathfinder* images were much crisper and they cost substantially less.

The most technically sophisticated piece of equipment on the

Viking lander was the biology package. It was remarkably small—a box about 12 inches on each side. The biology package weighed 33 pounds and contained 40,000 parts. In the original design, it contained four separate experiments that could be performed as many as four times. To conduct the experiments, the package contained tiny ovens, ampoules with nutrients inside, bottled radioactive gases, Geiger counters, some 50 valves, about 20,000 transistors, and a xenon lamp designed to duplicate the light of the sun. NASA officials originally estimated that they could develop the biology package for $11.3 million. The actual cost was $59.5 million.

The biology package could not fail. The search for living organisms in the Martian soil was the whole purpose for placing the *Viking* lander on Mars. Jim Martin, the overall Viking project manager, issued a 1971 directive stating that "no single malfunction shall cause the loss of data return from more than one scientific investigation." In other words, if some part of the biology package bombed, it could take only one experiment with it. This created a technical nightmare. The only way to pack four experiments into a small, lightweight container was to make the package complex. In an instrument this complex, any single-point failure threatened the whole system.[13]

The people developing the biology package were under tremendous pressure to simultaneously reduce cost, weight, and complexity. With four experiments, this could not be done. NASA executives thus confronted the sole solution to their dilemma: drop one of the experiments. In 1972, over the strong objection of scientists on the biology instrument team, NASA officials eliminated the "Wolf Trap" experiment. Named after its designer, Wolf Vishniac, the experiment would have tested the reaction of Martian soil to an aqueous solution. If microorganisms grew, they would make the liquid cloudy, which light sensors would detect.

The experiment disappeared. Cost growth continued. To compound the agony, Vishniac slipped and fell on a steep slope in an Antarctica dry valley that provided the most Mars-like environment on Earth. An expert on the ability of organisms to grow in hostile environments, he died alone traversing the valley that had inspired his dream of exploring Mars.

The team developing the *Sojourner* micro-rover faced challenges similar to those encountered by the Viking biology team. The rover team had to develop a small, semi-intelligent robot, about 2 feet long and a foot high that weighed no more than 22 pounds. The rover had to be smart enough to choose its own path across Martian terrain to targets selected by human operators in Pasadena, California. It carried a small alpha proton X-ray spectrometer which, when pressed against rocks or soil, could measure their chemical composition. It carried its own cameras. It had to communicate through the lander, its only pathway to Earth. It had to operate on the modest amount of power generated by a 1.9 square foot solar panel. At peak power, with the sun shining directly down, the panels generated 16 watts. Most of the time the rover operated on half of that. "Eight watts," rover manager Donna Shirley observed, "is about the power of the night-light in your bathroom."[14]

The easiest way to build such a rover and hold down its cost was to tether it to the lander. With a tether, the rover would not need its own power, radio, or brain. Shirley, however, did not want to build a rover on a leash. It would prove nothing technologically, and she knew that once the rover landed on Mars, scientists would ache to touch rocks that only a free-ranging rover could reach.

Shirley promised to build a free-ranging rover and stay within NASA's $25 million cap. To avoid cost overruns, she relied heavily on existing technology. For their cameras, team members used imaging chips found in home video cameras. The rover talked to the lander over a commercial Motorola radio modem that cost about $700. The onboard computer used a standard Intel 80C85 processor. For the engines that drove the rover's six small wheels, the team turned to a standard $100 motor manufactured by the Swiss firm Maxon. In each case, equipment had to be modified to work in the radiation-soaked, subzero environment of Mars. Modifications and subsequent testing were expensive. The basic technology was not.

Existing technology in the 1990s helped the Sojourner team develop the free-ranging rover for $25 million without cost overruns. This did not happen with the Viking biology package. The biology package started at $11.3 million—a reasonable sum given the instrument's capability—then soared to $59.5 million. Had Sojourner team members experienced similar overruns, it would have doomed their project. Much of the ability of the Sojourner team to avoid such overruns was due to advances in technology that had occurred in the intervening years.

This story was repeated frequently on the two landers. The computer on the *Viking* lander weighed 52 pounds. Its cost grew from $3.4 to $28.1 million. The Pathfinder team used a commercially available IMB RAD 6000. The *Viking* lander required 17 times as much electric power as the *Pathfinder* lander. That ruled out the use of solar panels, which the Pathfinder team used, and required the Viking team to install two 30 pound radioisotope thermoelectric generators powered by plutonium-239. The Earth-to-Mars radio on the *Viking* lander weighed 63 pounds.

No project team that attempts to develop a sophisticated space probe avoids technical challenges. Technical difficulties resulted in substantial cost overruns on the Viking project. On Mars Pathfinder, they did not. The reason, in part, was the state of technology. The Mars Pathfinder team had the advantage of 20 years of technological advances in miniaturization and micro-electronics.

Cost overruns on the Viking landers totaled $194 million.[15] Adjusted for inflation, that amounts to $716 million. Much of that savings was due to differences in technology.

FEWER, FASTER The Mars Pathfinder team also saved money by employing fewer people. The number of people required to manage any spaceflight expedition increases exponentially with the scale of the undertaking. A handful of people can oversee the development of a small spacecraft, but thousands are required to fly large systems like the space shuttle or trips to the moon.

The Viking project, at the height of its activity, employed 538 people at the Langley Research Center, 1,650 people at the Martin Marietta Aerospace company, an unknown number of people at the Jet Propulsion Laboratory, and 69 scientists on 13 advisory teams. The overall project management team, led by Jim Martin, was located at the Langley Center. The Langley Center was responsible for developing the *Viking* landers, a task it contracted out to Martin Marietta Aerospace in Denver, Colorado. The Jet Propulsion Laboratory was responsible for developing the orbiters and controlling the mission once under way.[16]

As more workers are added to a project, and more sites get involved in its development, additional people must be hired to coordinate them. The Pathfinder team avoided this trend by concentrating project management at a single center: the Jet Propulsion Laboratory. The central management team never comprised more than about 30 people. At the height of the undertaking, about 300 JPL employees labored on various aspects of the project. Another two dozen worked on the separately funded Sojourner team, which was also located at the Jet Propulsion Laboratory. The lander and *Sojourner* rover were built in-house at JPL and thus no single contractor was retained to oversee spacecraft development. The science advisory teams were kept small.

In keeping with the philosophy that projects should be "faster" as well as "cheaper," the Pathfinder team developed their spacecraft in 3 years, about half the time that the Viking team took. The Pathfinder team received its first funding in the fall of 1993; it launched its spacecraft in December 1996. Documents approving the Viking project were signed in February 1969; the launch of the two spacecraft took place in August and September 1975.[17] In sum, the Pathfinder team completed their project with about one-third of the number of people working about one-half of the time.

This created savings that reverberated throughout the project. People are the most expensive part of any spaceflight endeavor. They cost far more than machinery or propellants. At the time of the

Pathfinder mission, the government spent on average $95,000 annually for each NASA employee, including salary, benefits, and overhead.[18] A project that employs 3 times as many people over twice as long will run up huge personnel expenses. These expenses will appear throughout the project, in flight system budgets and science instrumentation as well as general management.

In general management alone, the Mars Pathfinder team saved $161 million. This activity includes project management, mission engineering and integration, and operational preparation prior to launch, expenses that are predominantly the salaries of people. In the dollars of their time, the Viking team spent $48.2 million on general management. The Pathfinder team, 20 years later, spent just $17.1.

OPERATIONS The effort by the Mars Pathfinder team to save money extended into spacecraft operations. The *Pathfinder* spacecraft was launched on December 4, 1996, and landed on July 4, 1997. The operations team was quite small—just a few dozen people. At times, according to team member Robert Manning, only a pair of people operated the spacecraft from the JPL mission control center. Operational expenses for the Pathfinder mission after launch totaled less than $15 million.

The Viking team, by contrast, assembled an army of experts to operate their spacecraft. Flight controllers and science teams flew the spacecraft, analyzed orbital photographs, debated landing sites, and conducted experiments on the ground. The Viking team spent $51 million on initial operations in the period immediately following the first landing. In inflation-adjusted dollars, that is $174 million more than spent by the Pathfinder team.

Surface operations on the Pathfinder mission were designed to last only a few months. The Viking project kept running for 5 more years. *Lander 1* was still operating, transmitting weather information and surface pictures when project funds ran out in 1980. NASA spent $51 million on "extended mission operations" from late 1976 through 1980, a sum that has been applied to extra capability.

The operations team at Pasadena lost touch with the *Pathfinder* lander on September 27, 1997, the 83rd day of surface operations. The *Sojourner* rover continued to rove, but without the lander's help, it could not communicate with its designers back on Earth. Every Martian day thereafter as the sun rose, the *Sojourner* rover automatically began a tiny minuet. It approached to within 3 meters of the mute lander, tried to establish radio contact, then moved back and circled the silent probe. With the onset of the Martian winter, the little rover grew cold and its electronic circuits eventually failed.

SUMMARY Members of the Mars Pathfinder team had to complete their mission for $3.6 billion less than what had been spent on the Viking mission 20 years earlier. Following the calculations outlined above, it is possible to summarize how this was done.

They cut launch costs by the equivalent of $240 million.

They did not include orbital spacecraft or as many scientific instruments as part of their project. They did not operate the *Pathfinder* spacecraft for a long period of time. Differences in capability saved the Pathfinder team about $1,038 million in inflation-adjusted dollars.

The Pathfinder team significantly reduced the cost of flight operations after launch. For the period of initial operations, this produced a $174 million savings. The remaining operational savings, created by limiting the length of the Pathfinder mission, have been assigned to reduced capability.

Members of the Pathfinder team employed far fewer people for much less time. In project management alone, that saved $161 million. Substantial savings due to a leaner workforce helped to cut the cost of scientific instrumentation and flight systems.

Pathfinder and Sojourner team members saved about $545 million on scientific instrumentation, exclusive of differences in capability. Much of that was due to their ability to avoid cost overruns on the equipment they developed, which was in turn traceable to better technology.

Finally, the Pathfinder team produced a single lander whose basic flight systems cost only $135 million. The flight systems for two *Viking* landers, in inflation-adjusted dollars, cost over $1.6 billion. The $1.5 billion savings were due to fewer people, less time, better technology, and a willingness to take greater risks.

ORGANIZATION

By the end of 1998, the year after the *Pathfinder* probe reached Mars, NASA officials had dispatched a sufficient number of "faster, better, cheaper" spacecraft for patterns to emerge. A report issued by the NASA Chief Engineer's Office, authored by Charles Cockrell, identified lessons learned and a succession of books written by participants in the Mars *Pathfinder* mission began to appear. These were succeeded, in the two years that followed, by management and accident reports occasioned by the missions that had failed.

In all, these reports revealed many of the practices that managers conducting "faster, better, cheaper" projects tended to adopt, as well as the practices they tended to avoid—sometimes to the detriment of the project. In general, managers conducting projects used organizational and technological reforms to cut costs. They then adopted a variety of risk management techniques to compensate for the reliability problems that lower spending and shorter schedules imposed.

With respect to organization, managers departed significantly from the methods historically used to develop complex space missions—full-scale systems management. Systems management arose as a means of developing projects utilizing risky technologies. Directors of early spaceflight missions adopted systems management because weaker forms of coordination failed; people who learn from failure do not readily relinquish those lessons simply because someone wants to save money. Project managers were not as comfortable with team-based coordinating techniques as with the older, more formal systems management methods. The transition from systems management to team-based organizations occasioned by "faster, better, cheaper" was a difficult one.

ORGANIZING FOR SPACEFLIGHT Since the beginning of the space age, experts have struggled to create workable institutions whose members can build rockets and spacecraft that actually fly. The history of spaceflight is littered with the remains of machines that failed because the people in charge did not know how to organize effectively. The lessons of high-technology management have been gathered from unforgiving errors.

In 1959, as an early step in the process leading to the landing on the moon, NASA executives established what became known as Project Ranger. The project was designed to return the first high-resolution pictures of the surface of the moon. NASA officials assigned the project to employees at the Jet Propulsion Laboratory in Pasadena, California. Build a spacecraft, NASA officials said, that could fly straight toward the moon, take close-up pictures, and transmit them back to Earth before crashing into the lunar surface. The project revealed how terribly disorganized the United States was for spaceflight and how difficult the management process would be.

JPL employees had already acquired a reputation for technical expertise by building high-altitude rockets and spacecraft. Established as a rocket development center for the U.S. Army, JPL housed the team that created the fourth-stage rocket that pushed the first U.S. satellite, *Explorer 1*, into orbit, as well as the satellite's mechanical and electronic systems. (James Van Allen of the State University of Iowa built the instruments.) JPL employees had also created the first U.S. space probe (*Pioneer 4*) to leave the gravitational field of the Earth and fly past the moon.

JPL is a strange organization. It is a NASA field center, but not officially part of the U.S. government. Instead, it functions as a contract organization, managed by the California Institute of Technology, one of the nation's premier scientific and technical universities. As an appendage of Cal Tech, it is organized along academic lines. It contains technical divisions, set up like university departments, with expertise in spacecraft engineering, navigation, telecommunications, and scientific instrumentation. In effect, JPL is a university that develops spacecraft instead of students. During the era of Project Ranger, the divisions were independent and very strong.

To conduct Project Ranger, JPL employees had to build a very complex spacecraft. Spacecraft developers employed solar panels, in-flight computers, small rocket motors, high-gain directional antennas, and special attitude stabilization systems. All of these were novel technologies for their time. Each spacecraft carried television cameras designed to take more than 4,000 television pictures as the spacecraft approached the moon and transmit them back to Earth as amplified electronic signals. To further complicate the mission, JPL employees had to mount their spacecraft on *Atlas-Agena* launch vehicles built by two separate industrial contractors working for the U.S. Air Force and managed by another NASA field center.

To oversee Project Ranger and similar activities, JPL executives established a special Lunar and Planetary Projects Office. People in the office utilized project management techniques already well developed by that time. These techniques had grown out of wartime development efforts like the Manhattan Project and Cold War activities like the Polaris missile submarine development project. In the form utilized at JPL, project managers assisted by a small staff drew upon people in the technical divisions to assemble experts who could design and build the *Ranger* spacecraft. The approach used at JPL is more appropriately known as matrix management—a variety of project organization in which most team members are temporarily detailed from technical divisions to serve on the project team.

The organizational scheme proved disastrous. The first six Ranger shots at the moon all failed. *Rangers 1* and *2* fell back to Earth when their *Agena* upper rocket stages failed. The spacecraft computers on *Rangers 3* and *4* failed. The power system on *Ranger 5* failed, accompanied by another computer failure. The television cameras on *Ranger 6* failed. Congress launched a major investigation, and NASA Administrator James Webb threatened to close JPL unless it adopted stronger management reforms. Not until the summer of 1964 did JPL employees carry out their first successful Ranger mission.

The Ranger mission demonstrated the limitations of a weak project organization. Matrix management was fine as a first step, but it did not sufficiently empower JPL's project managers with the tools they needed to successfully manage complexity. Project managers had to plead for scientific and engineering help from the technical divisions, which division chiefs could refuse to provide. Division chiefs could transfer personnel without the project managers' consent and they could ignore project management decisions. With only a small staff, JPL project managers were forced to rely upon division chiefs and ad hoc committees to integrate spacecraft components, solve technical problems, and conduct design reviews. The work of JPL's project managers was essentially reduced to that of tracking the spacecraft development schedule.

External relations were worse. Executives at NASA headquarters micro-managed the project, while Air Force officers charged with delivering *Atlas-Agena* launch vehicles ignored it. The contractor manufacturing the *Agena* upper stage gave far higher priority to the hundreds of rockets being manufactured for the Air Force than to the nine being prepared for NASA. When JPL managers asked for information needed to check "interfaces" between the *Ranger* spacecraft and the *Agena* upper stage on which it sat, Air Force officers replied that the managers did not have a "need to know."[1] When the contractor finally sent a mock-up of the *Agena* upper stage to JPL, project workers discovered that it did not match the *Ranger* probe.

The management problems that NASA officials experienced with the relatively simple *Ranger* probe were magnified tenfold in the crash program to put humans on the moon. Once again, NASA executives instituted project management, but failed to give project leaders the necessary tools to manage the mission. One of the most glaring omissions occurred in the relationship between project directors and the NASA field centers. Government officials established three NASA field centers to conduct the race to the moon: the Marshall, Kennedy, and Johnson Space Centers (the latter then known as the Manned Spacecraft Center). NASA executives established a program office at agency headquarters in Washington, D.C., to oversee the moon race, but did not force the Marshall, Kennedy, and Johnson center directors to report to it.

As was the case at JPL, the three field centers and their technical divisions were strong and independent. They behaved much like rival universities. The people who ran the technical divisions in the field had very little experience managing large projects or overseeing contractors. Their approach to project coordination, in the words of one observer, consisted of "management by committee." Project managers were forced to rely upon a "bond of mutual purpose" as committees met to solve design and integration problems.[2] Headquarters officials created committees to coordinate committees.

The event that precipitated an end to this loose system of organization was not a technical failure, although technical problems plagued the early human spaceflight program. The precipitating event in this case was a series of cost overruns on the Gemini project. Project Gemini was established in 1961 shortly after President Kennedy's decision to send Americans to the moon as a means of perfecting orbital rendezvous and docking techniques. Without successful completion of Project Gemini, astronauts returning from the moon would not have been able to reach the orbiting command module waiting to take them home.

NASA was barely 3 years old when Project Gemini began. Project workers had retained many of the organizational habits developed within the National Advisory Committee for Aeronautics (NACA), the agency from which NASA was formed. NACA contained many talented scientists and engineers, but none had experience managing large programs. Only a few people in the United States at the time understood the processes involved in managing very large and technically complex undertakings. NASA's weak management system for Project Gemini combined with growing technical problems to produce breathtaking cost overruns. NASA's estimated cost for completing the Gemini mission tripled in just two years.

When NASA's chief for human spaceflight requested more money for Project Gemini from James Webb, the NASA administrator, Webb told him no. The project's shortcomings were due to bad management, Webb replied, not money. When the flight chief appealed to President Kennedy, Webb replaced him and brought in a new management team.

The reforms instituted by the new team revolutionized the way in which government and industry managed spaceflight activities. The reforms spread quickly into other high-technology undertakings in business and government. In making spaceflight possible, the organizational reforms were as important as the technological innovations that they sought to control.

FORMAL SYSTEMS MANAGEMENT To institute the needed reforms, Webb turned to the U.S. Air Force. Previously, Air Force officers had contributed rocket technology to the NASA space program. John Glenn rode into orbit on a modified *Atlas* intercontinental ballistic missile (ICBM), and the first Gemini astronauts were scheduled to ride an Air Force *Titan II* when that flight program began. Webb asked the people who had developed those rockets to contribute their management methods as well.

During the decade before the moon race began, Air Force officers and their industrial contractors developed a number of management reforms which they used to complete the Cold War crash program to deploy a fleet of ICBMs. Ballistic missiles are highly integrated systems in which propulsion chambers, propellant units, turbo pumps (for liquid fuels), inertial guidance subsystems, in-flight computers, and electronic components must work in concert for any flight to succeed. The design of each missile subsystem affects the performance of the rest. Driven by the knowledge that the Soviet Union was developing its own ICBM fleet, Air Force officers had to develop missile components in very short periods of time. Responding to the pressures of the Cold War, military leaders formulated the concept of "concurrency." Instead of developing individual flight subsystems, production tools, ground equipment, and operational facilities in a sequential fashion, military officers ordered that all be completed simultaneously. The science of rocketry being primitive, design changes occurred frequently. Changes reverberated throughout the project, necessitating modifications in areas other than the component being changed. Not surprisingly, design changes produced cost overruns and, in the absence of any overall method for coordinating change, frequent technical failures. Missiles and rockets exploded with statistical regularity, sometimes destroying their test facilities. Experimenting with better methods for managing costs and controlling performance, project officers discovered what became known as systems management. Similar organizational reforms arose from the related effort to build a centralized air defense command system, an equally challenging task given the primitive state of computer technology at the time.

Air Force Brigadier General Bernard Schriever, charged with the ICBM development project, instituted most of the reforms. As project specialists had done before him, Schriever drew his subsystem managers into a centralized "weapon system project office," removed from day-to-day military operations and logistics. Contractors were told to organize themselves in a similar fashion. Schriever and his associates gave project managers a set of tools designed to improve their ability to integrate weapons subsystems. He and his associates, for example, insisted upon integrated systems testing, in which subsystems were placed in simulators that replicated operations of the system as a whole under field conditions.

The most important reform that Schriever introduced was known as configuration management. Schriever and his first-line managers discovered what they considered the best way to coordinate the work of project specialists. They merged the planning work that took place around cost and schedule with the engineering processes used to review design changes. As Stephen Johnson explains: "Because engineers needed to coordinate changes among themselves . . . they aired technical details of changes in coordination meetings. Recognizing this, managers inserted themselves into these meetings to understand what was happening, and soon required the engineers to give cost and schedule estimates for these changes." The resulting reforms created an elaborate, highly structured process for reviewing cost, schedule, and technical details at progressive points in the development of new weapons systems. Schriever tied engineering changes to budget plans in such a way that project specialists had to produce relevant paperwork or they would not receive approval to spend necessary funds. "Configuration management," Johnson observes, "became the link between engineering coordination and managerial control."[3]

Beyond a cursory commitment to project management, NASA officials had not adopted any such reforms. Webb, who was well-informed about management trends and eventually served as president of the American Society for Public Administration, knew that the NASA mission would not succeed without them. He asked Air Force General Samuel C. Phillips, a key player in the Air Force Ballistic Systems Division, to take charge of Project Apollo. Webb pressured George Mueller (pronounced "Miller") to become the associate administrator for all human spaceflight activities, including Project Apollo. Mueller had worked on the Air Force ICBM project as an executive at Space Technology Laboratories, an independent contractor. Air Force officers and the contractors with whom they had worked appeared throughout the NASA organization. One took over development of the *Apollo* spacecraft; another appeared as the chief operational officer at the Jet Propulsion Laboratory.

The work they did revolutionized the civil space program. NASA's ability to fly to the moon and complete the other great achievements of the early years was due to a merging of technical capability that already existed within the agency with management reforms imported from the U.S. Air Force.

At Washington headquarters, Mueller reorganized NASA so that the directors of the three field centers conducting human space-flight reported directly to him. Phillips set up a large office at NASA headquarters estimated to contain more than 1,000 civil service and contract employees who conducted systems engineering and integration tasks for Project Apollo. The reforms affected other NASA activities as well. Webb insisted that executives at the Jet Propulsion Laboratory, for example, accept the recommendations of people within the center who were agitating for the adoption of systems management reforms.

Project managers employing systems management techniques instituted design reviews and progressive design freezes. These are formal procedures by which rocket and spacecraft plans are reviewed and fixed at ever-greater levels of specificity as development work progresses. Once fixed, a component can be altered only by filing a special document that requires top-level approval and often has to pass through a special design change board.

Project managers set up problem identification procedures, or "P-lists." Subsystem managers met periodically with project directors to review the status of various spacecraft and rocket components, identify and rank problems (often by color, with red being the worst), and assign responsibility for resolving them.

Project managers set up their own systems engineering units. Systems engineering specialists used configuration control techniques to coordinate the work of subsystem managers. To check the checkers, and make sure that components were fabricated according to design, project managers set up reliability and quality assurance units. They penetrated contractor operations with on-site inspectors. They wrote formal documents that governed the work of specialists brought in from the old technical divisions. They insisted on formal documentation of all changes, test results, and fabrication details so that faults could be identified.

When NASA first got started, missions were coordinated by loose groups of technical specialists meeting together to discuss project details, often assisted by a small project management team. This is the way that Project Mercury was conducted, and it was the preferred method of operation at JPL. Technical failures and cost overruns forced NASA to adopt stronger structural reforms. The reforms not only changed the way in which projects were managed, they also shifted the locus of power. Control over project details shifted from scientists and engineers within technical divisions to teams of managers armed with formal planning and control techniques. This is the same shift that had been under way ever since Frederick Taylor introduced the notion that management specialists using statistical methods could calculate the "one best way" of performing each job.

The management reforms produced high levels of reliability. They made possible the exploration of the moon, by both humans and robots like the *Ranger* spacecraft. According to a number of commentators, the reforms permitted the rapid expansion of the American aerospace and computer industries and fueled U.S. economic growth in the second half of the twentieth century. The management reforms were adopted worldwide by technology organizations that could not otherwise compete with the United States.

The reforms came with a price, however. They required large staffs and ponderous procedures that were expensive to operate. To justify expensive procedures, projects had to be large enough for the reforms to be cost-effective. Larger projects required more elaborate procedures. As procedures became more elaborate, projects grew larger.

At the height of Cold War competition, ballooning cost and size did not seem to matter. NASA and Air Force executives adopted cultures of competence that put project performance well above concerns over cost and size. When the easy money days disappeared, aerospace executives found themselves with an intractable problem. All of their experience told them that projects conducted without formal systems management procedures would fail. The new reality of post–Cold War politics suggested that politicians would not fund big projects governed by elaborate systems management procedures.

Somehow the people who designed space missions had to fashion an alternative to large-scale, formal systems management that did not regress to the old days of project disorganization and failure. In the beginning, few people thought that this could be done. "Faster, better, cheaper" became the primary means of testing whether alternatives to formal systems management could be made to work.

THE TEAMWORK APPROACH Spacecraft managers use systems management because they want to reduce the likelihood that a particular mission will fail. Interactive failures, in which a change in one component creates difficulties with the operation of another, are the most serious sources of failure and the ones that formal systems management is best equipped to handle. People using systems management combat interactive failures through an elaborate tracking system whereby the ramifications of any single change are considered for the mission as a whole. A secondary purpose of formal systems management is cost and schedule control. To that end, engineering decisions are reviewed not only for their effect on other subsystems, but for their cost and schedule ramifications as well.

The people who manage "faster, better, cheaper" projects have exactly the same objectives: prevent mission failure while controlling cost and schedule. They attempt to do this, however, in a fundamentally different way. They cannot rely upon elaborate systems management procedures, which are too expensive and time-consuming. They do use elements of systems management, but given the very low costs and short schedules they are trying to achieve, these managers cannot afford to make systems management the cornerstone of their efforts to control cost, schedule, and risk. Instead, people who run such missions turn to the dynamics that arise in small, cohesive project teams.

For some time, management specialists have understood the latent force existing in small work teams. The phenomenon of small group productivity was first observed systematically during the Hawthorne Works studies at the Western Electric Company in Chicago and Cicero, Illinois. Scientists armed with the best structural management techniques of the day arrived at the Hawthorne Works in 1923, only to discover that their efforts to improve worker productivity through scientific management were confounded by the behavior of work teams. In one celebrated experiment, researchers attempted to improve productivity by scientifically adjusting the level of illumination in a workroom. Productivity increased when the research team improved the lighting; it increased when the research team made the lighting worse; and it increased when the research team set up a control group that received no new lighting at all. Repeated experimentation revealed that workers paid far more attention to the presence of researchers and their participatory methods of supervision than to objective factors like the level of lighting or the frequency of work breaks. The Hawthorne studies clearly confirmed that teamwork and supervision were critical factors affecting productivity in the workplace.

In the aerospace industry, the power of teamwork was demonstrated through the famous Kelly Johnson "skunk works." Clarence "Kelly" Johnson was an aircraft engineer who set up a special unit within the Lockheed corporation during World War II for the purpose of designing America's first production-line jet aircraft. Instead of organizing his unit in the conventional manner, with separate divisions and technical specialists working independently, Johnson assembled a small team of the best people he could find and isolated them from the main Lockheed plant. In the beginning, Johnson's team worked under a circus tent. According to one version, the term "skunk works" arose because the tent was located next to a plastics factory that gave off a distinctive smell. Johnson's skunk works philosophy was frequently employed thereafter in the aerospace industry, especially during the initial design phase of complicated aircraft that had to be completed in very short periods of time. Skunk works teams designed the U-2 spy plane, the SR-71 *Blackbird*, and the F-117A *Stealth Fighter*.

Skunk works philosophy is based on the recognition that a very small team of talented people can achieve high levels of productivity if they are given nearly complete control over their own work. Skunk works teams are kept deliberately small—10 to 25 percent of the normal workforce used for development projects. Smaller teams allow for more rapid communication and problem solving among members. Team members are isolated from the rest of the organization; they are given the authority to make design changes, select subcontractors, revise pay scales, manage their own budgets, and avoid excessive paperwork. They are expected to test their product extensively. Oversight is based on trust rather than formal procedures, which means that the team is pretty much left alone to accomplish its mission. A 14-point skunk works philosophy has received wide circulation in aerospace and business circles. (See box 4.)

The skunk works philosophy of small, cohesive work teams has been applied extensively in technology industries where high levels of creativity are required for product development. The computer software firm Microsoft began as a small, team-based organization. Microsoft executives Bill Gates and Paul Allen found that small work teams could design software programs in one-tenth the standard industry time. The software they prepared for ponderous giants like IBM allowed Microsoft to flourish during its infant years. Visual effects firms such as Digital Domain, which produced special effects for the movie *Apollo 13,* use skunk works-type teams. The aviation industry has used skunk works teams to develop commercial aircraft such as the *Boeing 777.*

In government, the value of small work teams is controversial. Government executives know that work teams endowed with high amounts of discretion can achieve impressive levels of creativity and service delivery. Granting discretion to enterprising work teams is the cornerstone of the effort to "reinvent government," the phrase used to describe the reform movement that swept through public management circles during the final two decades of the twentieth century. At the same time, many people fear excessive discretion in the hands of government employees. In government, an elaborate

1. The skunk works manager must be delegated practically complete control of his program. The manager should report to a division president or higher.

2. Strong but small project offices must be used by everyone involved.

3. The number of people having any connection with the project must be restricted in an almost vicious manner. Use a small number of good people.

4. A very simple drawing and drawing release system with great flexibility for making changes must be provided in order to recover from failures.

5. There must be a minimum number of reports required, but important work must be recorded thoroughly.

6. There must be a monthly cost review covering not only what has been spent and committed but also projected costs to the conclusion of the program. Don't surprise the customer with sudden overruns.

7. Team members must be delegated and must assume more than normal responsibility to get good vendor bids for subcontracts on the project.

8. Push basic inspection responsibility back to subcontractors and vendors.

9. Team members must be delegated the authority to test their final product in flight. They can and must test it in the initial stages.

10. The specifications applying to hardware must be agreed to in advance of contracting.

11. Project funding must be timely so that the skunk works manager doesn't have to keep running to the bank to manage cash flow.

12. There must be mutual trust between the customer and the skunk works team with very close cooperation and liaison on a day-to-day basis. This cuts down misunderstanding and correspondence to an absolute minimum.

13. Access by outsiders to the project and its personnel must be strictly controlled.

14. Because only a few good people will be used, ways must be provided to pay people based on performance and not on the number of people supervised.

Source: Lockheed Martin home page

system of legislative oversight and administrative law exists to discourage zealous behavior among public executives. In the minds of many people, excessive discretion is an invitation to the abuse of power.

Many of the practices that NASA executives adopted as the agency matured discouraged spacecraft development through small work teams. Projects grew larger, necessitating more elaborate methods of coordination and oversight. With fewer projects being approved, more people got involved in the ones that remained. Center directors lobbied for multicenter missions as a means of keeping their workforces employed. As missions grew more complex, and the participation of field centers proliferated, the need for formal systems management as a coordinating mechanism increased.

ENCOURAGING TEAMWORK Small, cohesive work teams are hard to maintain in large, ponderous bureaucracies. Many organizational forces conspire to defeat them. This is as true in industrial bureaucracies as in government ones. Managers of "faster, better, cheaper" projects used a number of techniques to strengthen their teams. Not all of the managers used all of the techniques, however; considerable variation occurred in practice.

One of the chief obstacles to the creation of cohesive work teams is the practice of distributing work to different field centers. NASA officials divided work on the Viking project to Mars, for example, between the Langley Research Center, which produced the spacecraft, and the Jet Propulsion Laboratory, which operated them. By contrast, officers concentrated the team-based Pathfinder mission wholly within the Jet Propulsion Laboratory.

Co-location of the project team at a single location is a principal technique for promoting teamwork. Not all of the "faster, better, cheaper" project managers employed co-location, however. Team members working on the NEAR project were not co-located at one site. They worked for a small organization—the Applied Physics Laboratory run by the Johns Hopkins University in Laurel, Maryland—that had a tradition of personal interaction on low-cost projects and access to electronic technologies that permitted close interaction in the absence of physical proximity. A NASA employee who reviewed the "faster, better, cheaper" projects, Charles Cockrell, observed that "close interaction" was the single most important factor contributing to project success.[4]

At the project site, overall responsibility for the mission was usually vested in a single core group. Following the skunk works philosophy, the groups were small enough so that members could communicate with each other easily and solve informally what otherwise needed to be resolved through formal procedures. While the groups typically contained people representing the major specialties necessary to complete the mission, they did not contain all of the people working on the mission. Spacecraft components typically were developed by subcontractors; specialists from technical divisions provided expert advice and helped test components.

Group cohesion is difficult to achieve in practice. The Mars *Pathfinder* group, whose members closely followed other team-building techniques, divided itself into two separate teams. An independent group at JPL headed by aerospace engineer Donna Shirley produced the *Sojourner* rover that accompanied the *Pathfinder* spacecraft to Mars. To promote cooperation, Shirley moved her office to the same location housing the lander team headed by Tony Spear. Relations between the two project leaders were contentious and, in Shirley's words, sometimes unprofessional. "A really creative team will probably be as contentious as it is brilliant," Shirley observed, trying to explain why her mission succeeded in spite of this schism.[5]

Project leaders fought to protect team members from excessive red tape. To build a spacecraft faster and cheaper, scientists and engineers sought to depart from conventional rules. "Brian [Muirhead] and Tony [Spear] were empowered to make those rule changes or just make up their own rules," said one member of the Mars Pathfinder team. "We were empowered to depart from that with a minimum of oversight."[6]

5 LESSONS LEARNED FROM "FASTER, BETTER, CHEAPER" PROGRAMS
Results of the Cockrell Study

1. Use cohesive technical teams with authority to do the job.	5. Use experienced personnel.
	6. Establish good communication.
2. Maintain visibility through reviews.	7. Conduct better up-front planning.
	8. Have clear requirement definition.
3. Use a design-to-cost philosophy.	9. Use technology with an appropriate readiness level.
4. Apply risk management techniques.	

Source: Charles E. Cockrell, "Lessons Learned from Better, Faster, Cheaper Concepts As Applied to Selected NASA Programs," Enterprise Safety and Mission Assurance Division, Office of Safety and Mission Assurance, NASA Headquarters, Washington, D.C., November 5, 1998.

Most core groups did not suffer from excessive outside interference, especially from above. This was especially significant on those missions where the core group was located outside NASA, working under government contract at sites such as the Applied Physics Laboratory for the NEAR project or the Orbital Sciences Corporation for the *Clark* satellite. NASA has a tradition of penetrating contractors and closely monitoring their work. NASA executives like to know when a project is going astray. This tendency was restrained on "faster, better, cheaper" projects, even when project teams got into trouble. NASA executives did not "penetrate" the Orbital Sciences Corporation even though the Clark project got into so much trouble that NASA eventually canceled it. Managers of the Mars exploration program at JPL did not penetrate the Lockheed Martin Astronautics team sufficiently to discover that its members were using the wrong units of measurement on Mars *Climate Orbiter*.

Protecting core groups from outside interference is a controversial principle. When a project fails, as did the *Clark* satellite and Mars *Climate Orbiter*, outside investigators typically recommend tighter oversight or better paperwork trails.[7] Ultimately, however, this defeats the teamwork necessary to make low-cost spacecraft work. Where a high-discretion project fails, it is often due to the violation of teamwork principles such as the need for co-location, rather than the absence of outside controls. Excessive oversight weakens the central team whose cohesion is necessary for mission success.

Not only did members of core groups attempt to protect themselves from red tape, they also sought freedom from the limitations imposed by the annual budget process. The federal government does not use a capital budget; nor does it approve lump-sum appropriations for entire projects. Instead, Congress appropriates funds year by year, a procedure that encourages bureaucrats to spend money they do not need or waste time waiting for appropriations they have not received. To develop low-cost projects, managers asked to receive funds when they were needed. Budget officers at NASA headquarters worked hard to protect these managers from the inefficiencies of the annual budget process. This was often accomplished by keeping projects so small and inexpensive that they fell below the scope of overall budget scrutiny. NASA executives can move funds on small projects in ways that are not permissible on large ones.

As noted above, not all of the people who worked on the project worked on the core team. The core group for Mars Pathfinder consisted of about 30 individuals, while some 300 people participated on the project during the period of most intense activity. The ability of the core group to rapidly enlist a supporting cast from the surrounding organization—and quickly dismiss them when their work was done—is a key feature of low-cost innovation. This requires a mind-set in which employees no longer view themselves as members of large bureaucracies operating under layers of supervision, but as entrepreneurs who come together to solve problems or accomplish specific tasks. It also helps if the people above those employees in the organizational hierarchy view themselves as transparent, which is to say that they do not interfere with the ability of specialists to support project teams. Without such attitudes, overall coordination is hard to achieve. The small scale of "faster, better, cheaper" projects, their short duration, and the fact that they contained capable specialists on the core team helped to improve relations with people from technical divisions whose assistance team members sought.

In a similar fashion, members of core teams sought to maintain close relations with contractors who provided spacecraft components. Contractors often organized themselves into cohesive groups that replicated the organizational style of the central team. The cameras on Mars *Polar Lander*, for example, were produced by the Lunar and Planetary Laboratory, an entrepreneurial group at the University of Arizona at Tucson. The laboratory has a reputation for producing precise spacecraft instruments, including the imaging systems for Mars *Pathfinder* and the Cassini mission to Saturn.

To encourage teamwork on the core group, many project leaders recruited a mixture of experienced and inexperienced personnel. Experienced personnel led the team; inexperienced personnel served in supporting roles. Senior personnel served as mentors to less experienced team members. Inexperienced personnel helped team members discover new methods for achieving project goals.

Team members actually worked with the spacecraft, a critical factor in the development of problem-solving capacity. In the aerospace community this is known as "hands-on" work. As simple as it sounds, hands-on activity is not always guaranteed. On large projects, spacecraft assembly and testing frequently take place at sites well removed from the central team. In such cases, project leaders produce paperwork, not spacecraft. The management philosophy underlying "faster, better, cheaper" reverses this trend. Assembly and testing of the whole spacecraft take place under the direct control of the central team. (Fabrication of individual components may be done by contractors.) Hands-on work motivates team members, encourages initiative, and engenders a sense that members are personally responsible for the success of the spacecraft.

Some team leaders built capacity through multitasking. This is the practice of asking core team members to perform more than one job as the project evolves. This is closely associated with the practice of seamless management. Under the latter feature, as people from the core team finish one task, they move on to new responsibilities. As a consequence, people who design and test the spacecraft at the beginning of the project become its operators once it flies. This enhances institutional memory and helps to keep the core group small.

An important element in the process of team building is the willingness of members to share common goals. Goal sharing provides the glue that holds teams together while allowing members to exercise a high degree of individual discretion. Successful project leaders spend a great deal of time communicating goals to team members and making sure that members accept them. One of the most important goals is cost control. "Ownership of the cost constraint permeated all the way down to the lowest level of the organization," one of the members of the Mars Pathfinder team observed. Team members learned that they could not solve a problem by pulling in more people or spending more money.[8]

Using approaches like multitasking and shared goals, project leaders sought to produce a setting in which individual members felt so responsible for the overall success of the mission that they would personally try to solve problems wherever they occurred. "You can't build in enough checks and balances to catch everything," observed one of the leaders of the Mars Pathfinder team. Formal paperwork systems with their divisions of responsibility are always imperfect. They will inevitably miss a critical detail that can cause the entire mission to fail. "And everybody knows it," the team leader said. "It's up to the integrity of each individual to ensure that their hardware and/or software will do what it needs to do when tested against the rigors of space. The working team has to trust that they can bring a problem to the managers without getting their heads taken off. And management has to have people that they can trust, people to whom they can give the responsibility and authority to make things happen."[9]

Creating this sense of shared responsibility is expedited on a team that is too small. Understaffing was a key component of the "faster, better, cheaper" philosophy. It helped to be understaffed, said one team member. "Having a team that was as closely knit as we were had benefits in that everybody could see what wasn't getting done. So people would just naturally move in and do things that needed to be done, where there was no obvious other person that could do it."[10]

It also helped to be invisible. NASA executives were preoccupied with developing the International Space Station and flying the space shuttle when the Discovery program of low-cost missions got started. Executives at the Jet Propulsion Laboratory were preoccupied with the Cassini mission to Saturn while the Pathfinder mission was being formed. Being small and relatively invisible helped "faster, better, cheaper" team members convince themselves that they really had the power to solve their own problems without excessive oversight.

The independence of these early projects was exemplified by Donna Shirley's effort to put a name on the little rover bound for Mars. With the help of the Planetary Society, Shirley asked students to suggest the names of women who had blazed previous trails. The winning entry was submitted by a 12-year-old Connecticut girl who suggested the name of the nineteenth-century abolitionist Sojourner Truth, but not before a bureaucrat at NASA headquarters discovered the contest and ordered Shirley to stop. NASA had a regulation setting out proper procedures for naming missions, whose tenets Shirley had violated.

Naming a spacecraft is not a mission-threatening event, but the resolution of the incident demonstrated the independence of the rover technical team. Shirley made a few insignificant changes in the contest procedures, NASA bureaucrats backed down, and *Sojourner* went to Mars.[11] At the time, the bureaucrats at NASA headquarters did not comprehended how wildly popular the little rover would become. Had they known this, they would never have granted Shirley the degree of technical independence that she enjoyed.

After *Pathfinder* landed on Mars, awareness of the "faster, better, cheaper" approach grew. Subsequent teams launching spacecraft to Mars did not enjoy the invisibility that the Pathfinder team had experienced in its pre-landing phase. Not surprisingly, the discretion afforded team members changed and two of the subsequent missions failed.

The level of technical discretion granted team members on early "faster, better, cheaper" projects was unusually high by government standards. Where taxpayer funds are involved, politicians and government executives often demand a degree of accountability that reduces technical discretion. Government officials refuse to leave contractors alone; they reconsider project goals. Separating implementation from policy formulation is a principle that looks good on paper but is rarely achieved in practice. Sustaining the high levels of discretion that characterized early "faster, better, cheaper" projects proved difficult as the approach matured.

While the techniques utilized on these low-cost projects emphasized team building, team leaders did not wholly ignore formal procedures. Project leaders engaged in formal up-front planning and clear requirements definition, as recommended by the Hearth Commission. Up-front planning means that goals, responsibilities, standards, and funding limits are clearly defined before detailed work begins and are not changed once the project is under way. As Hearth observed more than a decade earlier, design changes and requirement drift inevitably lead to increased cost and spacecraft complexity. They also invite outside intervention of the sort that compromises the sense that team members can handle project details.

Some team members made use of the central recommendation of the Low Cost Systems Office, although most were too young to remember that George Low had offered it. Team members relied upon commercially available components or parts that had been flight-proven on previous missions. The Mars Pathfinder team, for example, saved time and money by using a high-altitude aeroshield and parachute subsystem taken from the 20-year-old Viking project and a spacecraft camera that had already been tested for the Cassini mission to Saturn.

Finally, project teams utilized formal systems management procedures. They conducted preliminary design reviews, critical design reviews, pre-ship reviews, flight readiness reviews, independent annual reviews, critical milestone reviews, monthly internal reviews, weekly internal reviews, problem identification reviews, and formal risk management. They conducted such reviews, but made sure that paperwork was not the guiding mechanism for project coordination.

Setting a balance between the use and overuse of formal reviews was accomplished in a number of ways. Team leaders avoided widely attended, "gala" reviews. By avoiding highly structured reviews, they encouraged members to work on project hardware instead of project paperwork. Team leaders made extensive use of peer reviews (also called "tabletop" reviews), in which people at the subsystem level present their design decisions to small groups of experts who are generally not members of the project team. Peers critiqued subsystem work and suggested changes, which members were free to ignore. Peer reviews were also used to select the projects that would qualify for overall funding.

The peer reviews were very powerful. "The peer reviews were the thing that made the project, that made the spacecraft work," said one team member. Formal reviews "were more for the management, to keep them happy." Formal reviews were viewed as useful for people outside the project, but not for people inside the team.[12]

PROJECT FAILURE When management problems arose, investigators often traced the difficulties to issues arising from the twin challenges of complexity and coordination. The principle of requisite variety, often cited by management specialists, states that any self-regulating system seeking to overcome the challenges posed by its environment must be as complex as the environment with which it deals.[13] Operating a risky technology in a hostile environment like space requires project organizations that are quite complex. Having established a complicated organization, managers who seek to control risky technologies must then coordinate it. The more complex the organization, the more powerful the methods of coordination must be.

As a number of experts have observed, the twin requirements of complexity and coordination tend to push organizations in different directions—especially where risky technologies are involved.[14] Risky technologies like spaceflight require very complex organizations and very powerful methods of coordination. The former create centrifugal forces that push organizations apart. To pull them together, centripetal forces by the laws of nature must be equal and opposite to outward force. People in complex organizations managing risky technologies as a result tend to pursue apparent opposites—they must simultaneously centralize and decentralize, delegating discretion to a widening array of specialists and concurrently pulling them together.

Sometimes a project organization is more complex than the mechanisms for coordinating it. In creating the project organization for the International Space Station, NASA executives divided responsibilities into "work packages." Each package was assigned to a different NASA field center, each assisted by its own collection of contractors. Anxious to expand their own development activities, center directors lobbied for a wide distribution. The distribution that emerged did not produce what engineers call "clean interfaces." Officials at the Glenn Research Center in Cleveland, Ohio, for example, lobbied to build the electric power generating system. Electric lines and equipment permeated the entire station, threatening to compound coordination problems with field centers designing other components. In the beginning, NASA executives worried that inclusion of additional field centers would produce a project too complex to coordinate. As one participant observed, complexity increases exponentially with the number of centers involved. The lines of communication among four field centers, for example, are twice as large as the lines among three.[15]

To broaden their base of support, NASA executives agreed to establish four major work packages—the number that the center directors preferred. The introduction of international partners from Europe, Canada, and Japan increased complexity even more, as did the addition of the Russian government in 1993.

As the complexity of the space station organization increased, so did the need for a strong project office that could coordinate all the parts. At the beginning of the program, however, field center directors opposed this. They lobbied for a relatively weak "lead center" approach in which station managers at one field center would coordinate workers at the rest. As proposed, members of the coordinating group would report to the center director at their installation, ensuring that the center directors would maintain the upper hand in space station development. The concept was wholly unworkable, but NASA headquarters executives accepted it because the center directors reached a consensus that it should be done. Shortly after President Ronald Reagan instructed NASA to proceed with the space station, NASA Administrator James Beggs designated the Johnson Space Center in Houston, Texas, as lead center for this complicated undertaking.[16]

The space station program was more complex than it needed to be. The coordinating mechanisms were weak. One Air Force executive who had set up systems management procedures for Project Apollo, called back to review these management practices, observed that "I was frankly surprised at the extent to which practices and disciplines that had been, I thought, pretty soundly established within NASA had fallen into disuse."[17]

Similar tendencies afflicted "faster, better, cheaper" projects. Corporate executives overseeing development of the *Lewis* satellite split the members of the managing team between East and West Coast offices. NASA executives overseeing Mars Climate Orbiter split spacecraft development responsibilities between an industry team in Denver, Colorado, and a Jet Propulsion Laboratory team at Pasadena, California.

NASA executives eventually established mechanisms to coordinate the space station development effort. They abandoned the weaker "lead center" approach and set up a strong systems management activity. On a well-funded project with a long development schedule, systems management provides a solution to the problems of complexity that risky technologies provide. It is solution with which spacecraft engineers are familiar and one that is part of the NASA tradition. Spacecraft managers conducting "faster, better, cheaper" projects cannot follow this path. Given the nature of their objectives, this is not a feasible strategy. Cost and schedule constraints (not enough time nor money) preclude the use of formal systems management. Instead, project managers are encouraged to substitute teamwork strategies such as those described above.

For many of the managers overseeing "faster, better, cheaper" projects, however, this proved difficult. The very factors that created complexity on their projects, such as the desire to proliferate project responsibility, interfered with the teamwork strategies necessary to overcome them. Centrifugal force exceeded centripetal strategies. This proved to be a perplexing difficulty. NASA's tradition of dividing up work and the relative unfamiliarity with team-based management worked to create an imbalance between complexity and coordination.

A few projects escaped this fate. On very small missions, pressures to distribute project responsibility did not prevail. On a few of the larger missions, such pressures did not arise. Members of the Mars Pathfinder team, for example, isolated themselves from the dominant NASA culture. Center executives were preoccupied with the Cassini mission to Saturn and the Pathfinder budget was too small to attract extensive oversight from executives at NASA headquarters.

Budgets for subsequent Mars projects were smaller still. The combined development budgets for the Mars Climate Orbiter and Polar Lander projects that failed in 1999 were significantly less than the development budgets for the Pathfinder and Global Surveyor projects that preceded them. Project workers spent $306 million to develop Pathfinder (excluding the *Sojourner* rover) and Global Surveyor, while the people working on the latter missions received only

$193 million. A number of observers pointed to budget reduction working in combination with already short development schedules as a principal cause of the coordination problems that arose. To them, the projects were too short and too cheap to succeed.

The analysis here suggests a slightly more complicated interpretation. Relative to the complexity of the mission, too little money and too little time can create circumstances in which project workers inadequately test or review their spacecraft. So too can an imbalance between project complexity and methods of coordination. Both forces afflicted the projects that failed.

In assessing the failure of three Mars missions in 1999, a special assessment team concluded that the "faster, better, cheaper" approach was "an effective concept . . . that should continue." They qualified that statement, however, by adding that the approach had to be "properly applied."[18] The relatively low costs and short development schedules that characterize such projects reduce the margins that managers of risky technologies commonly employ to conquer problems of reliability. Absent those margins, managers cannot afford the excessive complexity or weak coordinating techniques that NASA traditions often provide.

TECHNOLOGY

Management reform provides one avenue for reducing the cost of spaceflight. Through management reform, project leaders can reduce the size of the teams needed to develop and fly spacecraft. Management reform does not take place in a vacuum, however. Smaller teams are aided in their work by smaller, less complex spacecraft. Reducing the size and complexity of spacecraft through technology change was as important to the implementation of the "faster, better, cheaper" initiative as management reform.

Advocates of the initiative maintain that spacefaring nations cannot afford large, *Cassini*-class spacecraft anymore. The *Cassini* spacecraft, set to arrive at the planet Saturn in the summer of 2004, is one of the most sophisticated robotic probes ever developed. It is as big as a 30-seat yellow school bus. It carries scientific instruments capable of conducting investigations in almost inconceivable ways, including a 770 pound Huygens probe that will parachute into the atmosphere of Titan and, hopefully, land.

The *Cassini* probe cost $3.3 billion to develop, launch, and operate ($660 million of that was contributed by European space agencies, major partners in the mission). *Cassini* took 8 years to design and develop—as long as it took for the United States to go to the moon. At 12,346 pounds (including fuel for braking at Saturn), it weighed so much that its *Titan IVB/Centaur* launch vehicle could not push it directly toward Saturn. Instead, the launch vehicle set the spacecraft on a circuitous route that allowed it to accelerate as it whipped by Venus twice, the Earth, and Jupiter over a flight lasting nearly 7 years.

If anything went wrong with the mission, the U.S. government stood to lose billions of dollars and years of work. To prevent that, spacecraft designers built in lots of redundancy and safety features. Engineers equipped *Cassini* with three electric generators, 82 heaters, two main engines, three antennas, two data recorders, two main computers, and eight gyros. No spacecraft had visited Saturn for 20 years, and no spacecraft that expensive would visit for decades hence. *Cassini*'s designers wanted to make sure that their mission would not fail.

As an alternative to one big spacecraft, project managers could have divided up the instruments and put them on smaller probes, launching one every few years. In theory, this might seem to be more efficient, since smaller spacecraft cost less than larger ones. It might moderate the consequences of failure, since loss of a single spacecraft would sink only part of the mission. It might allow project managers to use more advanced technology, since they would not freeze the design for latter spacecraft until the first ones flew.

In fact, smaller spacecraft are not inherently more efficient. The practice of flying the same instruments on smaller spacecraft does not save money. Project managers must pay for more spacecraft, more computers, more tracking, and more data transmittal time to achieve the same results. Overall, smaller spacecraft are not automatically cheaper than larger ones. "If you just naively break that up into five missions," said one NASA scientist looking back at Cassini, "it's not going to be cheaper, it's going to be much more expensive."[1]

To reduce size in a cost-effective way, aerospace engineers experimented with new methods of building spacecraft. They worked to alter the technology on which spaceflight depends. They pursued fundamental changes in technology that altered spacecraft design in the following ways:

▲ Spacecraft designers sought to develop scientific instruments and spacecraft components dramatically smaller and easier to operate.
▲ They sought to develop methods of propulsion far more efficient, plus new methods of slowing and landing spacecraft once they arrive at their destination.
▲ They worked to develop smarter spacecraft capable of operating more autonomously with less ground control.

Spaceflight visionaries dream of the day when spacecraft become so small and cheap that practically anyone can build them. If this happens, humans will be able to flood the solar system with small, inexpensive space probes. Machines weighing a few pounds with the capabilities of previously mammoth spacecraft would provide a virtual human presence throughout the solar system, informing any person with access to a computer of the weather on Mars or the composition of the latest comet. Along with new methods of management, technology improvement is the broadest path to that goal.

COMPONENT TECHNOLOGY "By 2010, second-generation microspacecraft the size of toaster ovens that weigh 5–1/2 kilograms (about 12 pounds) and use 5 watts of power will travel a billion miles away and send data back to Earth." So claims a publication from NASA's Center for Space Microelectronics, located at the Jet Propulsion Laboratory in California. "Many of the key technologies will be derived from those in such commercial products as cell phones, low-power palm top computers and pagers."[2]

Using principles from the micro-electronics and computer industries, advocates of "faster, better, cheaper" hope to make this possible.

Begin with a device as simple as a tape recorder. Spacecraft like the *Hubble Space Telescope* flew in the 1990s with reel-to-reel tape recorders, used to store scientific data for later transmission back to Earth. Although custom-designed for the rigors of space, these were conventional recorders, with movable parts and tape that wound from reel to reel. They were also one of the most failure-prone items on modern spacecraft. When spacecraft designers began to search through the micro-electronics industry for new ideas, one of the first places they stopped was in the digital domain. Old-fashioned tape recorders work by recording electronic signals on a magnetized tape. Computers, scanners, fax machines, the Internet, compact disk players, and bar code readers all work by converting images into numbers. In its modern form, digital information can be stored on memory chips. Such chips are small, use less energy, and have no moving parts. In 1997, astronauts visited the *Hubble Space Telescope*, removed the old reel-to-reel machines, and substituted solid-state recorders. The use of solid-state recorders is a prime example of the way aerospace engineers utilize technology from the micro-electronics and computer industries to alter spacecraft design.

Some aerospace engineers borrow new technologies; others invent them. In some cases, aerospace engineers develop technologies that reshape industrial practices. Many spacecraft carry cameras. The first satellites to take pictures from space used actual film, which had to be dropped from satellites, retrieved by aircraft, and processed in a professional lab. This technology was employed for the Corona reconnaissance satellite program set up to spy on the Soviet Union during the Cold War. As technology improved, spacecraft designers installed charge-coupled devices (CCDs). A typical CCD contains hundreds of tiny light detectors. When an image falls on a CCD, each detector produces an electric signal measuring the intensity of the light reaching it. The signal can be measured digitally. CCD spacecraft technology is similar, although far more sophisticated, to that used to produce camcorders and digital cameras. A CCD was used for the camera on the Mars *Pathfinder* lander. It was taken from excess parts produced for the camera on the *Cassini* probe.

Charge-coupled devices are a nice technology, but spacecraft designers want to do better. Scientists and engineers want to replace CCDs with what they call "cameras on a chip." The camera for Mars *Pathfinder* was about 12 inches long, and was full of mirrors and gears and lenses. By employing active pixel sensors, scientists hope to build spacecraft cameras no larger than the size of a plastic gambling die. The cameras would be able to detect ultraviolet, visible, and near-infrared light. They would automatically adjust their exposure and electronically pan and zoom. A special camera would contain its own power source and transmitter, all in a one inch cube. Such cameras could be mounted on rovers or scattered throughout space.[3]

The camera technology is similar to that found in CCDs in that a "camera on a chip" would employ an array of light-sensitive pixels. On an active pixel sensor, however, each pixel has its own amplifier and transmitting line. The array operates much more like a computer chip than a conventional CCD. The controls and equipment that must be separated from a CCD could be incorporated into the active pixel sensor. All that the active pixel sensor would need is a small power source and a lens to focus light on it.

In the mid-1990s, a design team at NASA's Jet Propulsion Laboratory took the lead in developing active pixel sensors. Applied to spacecraft, the technology would cost less, weigh less, and use far less power. It would find use on scientific and military spacecraft, and have huge applications in the consumer marketplace.

Here is another example. Nearly all spacecraft depend upon gyros for navigation and control. A traditional gyro, or gyroscope, depends upon the stability inherent in a rotating mass. The axle of a spinning gyro will remain pointed at the initial direction to which it is set. Placed within an elaborate network of gears and sensors, a spinning gyro can be used to calculate changes in spacecraft direction; working in combination with accelerometers that sense changes in velocity, it can track the spacecraft's position with a high degree of accuracy. The mechanical gyros commonly used in spacecraft are bulky, power-hungry, delicate devices whose many moving parts invite failure. Spacecraft designers have long dreamed of gyros that have no moving parts. Given that gyros traditionally depend upon spinning wheels to create inertia, this long seemed an impossibility.

Both NEAR, the first Discovery program spacecraft, and the

Cassini probe were fitted with gyro systems with no moving parts. Both spacecraft relied upon a new technology called hemispherical resonator gyros. This technology, developed in the electronics industry, uses a resonating component to determine attitude. It measures the force required to rebalance a standing wave pattern on a quartz hemispherical resonator.

By nature, engineers like to tinker and improve. The people who promoted "faster, better, cheaper" knew that spacecraft engineers would install new technologies if they were encouraged to do so. Many of those technologies already existed in the micro-electronics and computing industries; others had been installed on larger spacecraft. Four of the new technologies used on low-cost spacecraft, for example, were pioneered by engineers working on the *Cassini* probe.

Managers were encouraged to adopt technologies that made their spacecraft smaller, cheaper, and more reliable. Testing new technologies was one of their mission goals. The people who built Mars *Pathfinder* experimented with new methods of landing; they also agreed to test micro-rover technology. NASA officials set up the New Millennium program specifically to flight-test new technologies. *Deep Space 1*, the first New Millennium spacecraft to fly, tested 12 technologies, including an ion propulsion engine, a miniature integrated camera spectrometer, and a "remote agent" computer software program that bordered on artificial intelligence. *Deep Space 2* flight-tested nine technologies, including an ultra-low-temperature lithium battery and a very-low-power micro-processor. Project managers are told to worry less about risk and more about advancing the state of spacecraft design. "If we fly it and it doesn't work but we know why," said one of the scientists working on the New Millennium program, "we're a success."[4]

PROPULSION TECHNOLOGY Changing component technology is an important step in spacecraft redesign. Most of the bulk that leaves the Earth for space, however, does not consist of scientific instruments or operating components. It consists of engines and fuel. Instruments and equipment sit on top of a pyramid of propellant. As any rocket scientist knows, improved propulsion technology would vastly expand the ability to work in space.

Here is how a typical interplanetary launch works. Mars *Polar Lander*, set to arrive in the last month of the old millennium, was launched on a *Delta 2*, model 7425. Three rocket stages were required to send the lander toward Mars. The first stage carried 316,000 pounds of fuel for a kerosene-burning engine and four solid rocket boosters. The first stage reached speeds of 12,000 miles per hour— insufficient even for low Earth orbit. The second stage contained 13,000 pounds of hydrazine and nitrogen tetroxide. It took over at an altitude of 76 miles and propelled the rocket into a low Earth orbit 119 miles high.

The *Delta 2-7425* is capable of delivering about 7,000 pounds to low Earth orbit. Still more fuel and engine mass is required for interplanetary flight. For the Mars polar lander, most of the weight that the *Delta* rocket delivered to low Earth orbit was consumed by a Thiokol *Star* 48B third-stage engine and about 4,400 pounds of solid rocket fuel. The *Star* 48B engine burned for 88 seconds, propelling the spacecraft from its parking orbit around Earth into a relatively slow, 11-month journey to Mars.

Sitting on top of this mass of rocketry and fuel, the Mars *Polar Lander* weighed 1,270 pounds. The actual lander, however, weighed about half of that. Spacecraft designers set aside the other half for the engines and power and braking equipment necessary to get the spacecraft onto Mars, including 141 pounds of fuel.

Suppose that you want to send a larger spacecraft to Mars with more capabilities. You could purchase a larger rocket, which would cost more, or you could use the weight currently devoted to the third stage in a more efficient way. The third stage, like the two before it, is mostly propellant. Every pound of propellant you can cut increases the mass of spacecraft and scientific instruments that can be flown.

A nuclear-powered engine consumes one-third of the propellant used by conventional chemical rockets. NASA studied nuclear-powered engines during the first decade of spaceflight, and tested one in 1969. Unlike fusion engines (still an imaginary form of propulsion), nuclear-powered engines do not use explosions to propel themselves. Rather, they employ a very hot nuclear core to increase the temperature of liquid hydrogen. The rapidly expanding hydrogen gas provides the thrust necessary to propel the accompanying spacecraft.

Nuclear-powered engines are fast and more efficient than chemical third stages. They are also environmentally and politically controversial. When "faster, better, cheaper" began, NASA officials did not experiment with nuclear-powered engines as a way of improving spacecraft capability. Instead, they experimented with ion propulsion, which is 10 times more efficient than conventional chemical engines from the standpoint of fuel and much less controversial.

Ion propulsion is actually solar-electric propulsion, so named because of its dependence on electricity generated through solar panels. It works by emitting ionized gas from an engine at very high speeds.

An ion propulsion engine was developed and flown on the first New Millennium spacecraft, *Deep Space 1*. The spacecraft was launched in October 1998, for a test drive around the solar system. The ion propulsion engine on *Deep Space 1* uses xenon, a rare gas that occurs naturally in the Earth's atmosphere. Mercury or cesium could be used as well, although both are more difficult to handle. Small amounts of gas are injected into the ion propulsion engine. A cathode, much like the one in a television picture tube, sits at the front of the engine, emitting electrons, which strike the xenon atoms as they enter. The free electrons knock away one of the 54 electrons contained in each xenon atom, ionizing the gas.

Having lost one of its electrons, the xenon gas carries a positive charge. A pair of electrically charged metal grids mounted across the engine's open end attract the ionized gas. The force of the electrical charge causes the xenon ions to shoot through the grids at speeds exceeding 60,000 miles per hour. An electrode at the rear of the engine injects more electrons into the exiting beam, neutralizing its positive charge in such a manner as to keep the atoms from being sucked back into the engine. The ionized exhaust emerges as a pale-blue beam. Counteraction propels the spacecraft forward with an almost imperceptible force.

The engine's advantage lies in its ability to deliver continuous thrust. The ion engine on *Deep Space 1*, a test model, measured just 12 inches in diameter and carried 181 pounds of xenon propellant, enough to last for 20 months. It drew its electric power from two solar panels just 63 inches long. If run for the full 20 months, the ion engine on *Deep Space 1* would have increased the spacecraft's speed by about 10,000 miles per hour. Increasing the velocity of a space-craft with a chemical propulsion engine would require 10 times as much fuel.

Launched on a *Delta 2*-7425, an ion propulsion engine theoret-ically would allow spacecraft designers to substitute 3,000 pounds of spacecraft for the rocket fuel carried on a conventional third-stage booster. In practice, designers might increase the spacecraft weight less than the full 3,000 pounds, and use the residual for a small chemical-fuel rocket to boost the spacecraft to a faster starting speed from which the ion propulsion engine would begin its work.

Additionally, anything that could be done to improve the first and second rocket stages would also improve mission efficiency. While *Deep Space 1* was flying around the solar system, both NASA and a variety of privately financed aerospace firms were experiment-ing with ways to cut the cost of first- and second-stage rockets. Officially, only the ion propulsion engine was part of the "faster, better, cheaper" technology improvement initiative.

The disadvantage of ion propulsion lies in its slow acceleration. Given the mechanics of spacecraft trajectories, however, that does not take the "faster" out of "faster, better, cheaper." A spacecraft bound for Pluto using a single burst of chemical propellants would take 10 to 12 years to glide to the outer limits of the solar system. An ion propulsion engine would accelerate slowly but steadily. It would need solar energy available only in the inner solar system. Month by month, it would cruise through the inner solar system, steadily picking up speed until it was ready to move out to regions where the sun would no longer power its solar arrays. Like the tortoise bypassing the hare, the ion-propelled probe would arrive at Pluto before the spacecraft powered by conventional chemical fuels.

Once a spacecraft arrives at its destination, it generally needs to slow down. Engineers designing the second New Millennium project approached this problem in an unconventional way. They attached two miniature probes to the interplanetary cruise equipment enclosing the Mars *Polar Lander*. The lander was designed to set down in a conventional way, using a chemical descent engine similar in concept to the ones designed for the *Viking* landers some 25 years earlier. When the lander moved away from its cruise stage, the tiny probes would separate and fall toward Mars unaided by any landing devices whatsoever.

Scientists calculated that the two probes, employing no parachutes or retro-rockets or inflatable air bags, would crash into the Martian surface at approximately 400 miles per hour. The instruments they contained had to survive this brutal blow. Indeed, scientists wanted the probe to hit hard, since part of its mission required the probe to propel a bullet-shaped object deep into the Martian soil and deploy a tool that would drill for evidence of water. Imagine trying to design a desktop computer with a printer on a 2-foot cable that would continue to operate after being dropped from an airplane.

The ability to hard-land miniature instruments on other bodies would vastly expand the capability to explore cheaply. NASA officials made hard-landing feasibility tests one of the principal technology goals of *Deep Space 2*. Approaching the problem, project engineers used what they called "an aggressive test program with a continuous design/develop/test/fix approach."[5]

To start their tests, engineers dropped a variety of objects out of airplanes flying 2 miles high. They threw out clay pots, Styrofoam cases, and Pyrex containers. The structures failed. Engineers quickly learned that cushions did not protect delicate instruments descending at high speeds. As an alternative, the engineers considered rigid structures in which shockwaves might flow through the structure without breaking internal components. They designed an outer shell made out of silicon carbide—the material in sandpaper. To test their concept, they shot prototypes out of an airgun at Elgin Air Force Base in Florida. The outer shell absorbed the impact, shattering on contact. The concept seemed to work, so they built the probes and sent them to Mars.

The spacecraft had to land right side up, of course, in order to point its antenna up and send its penetrator probe down. That turned out to be a relatively easy problem to solve. The engineers shaped the *Deep Space 2* aeroshell in the form of a badminton shuttlecock. Even if it tumbled at separation, it would turn upright soon after entering the atmosphere.

In the first 10 years of "faster, better, cheaper," scientists and engineers tested two new methods of landing spacecraft on another body: air bags with Mars *Pathfinder* and direct impact with *Deep Space 2*. They tested ion propulsion technology. By themselves, these were not particularly new technologies. Air bags were already in use on automobiles; engineers at NASA's Glenn Research Center had built an ion propulsion engine in 1959. Transforming concepts into spacecraft equipment and risking missions by flying them in space— that was essentially new. With risk came the possibility of failure. Air bags and ion propulsion worked; the *Deep Space 2* aeroshells did not, the consequence of what investigators concluded to be an inadequate test program.[6]

AUTONOMOUS SPACECRAFT The cost of operating spacecraft once launched is substantial. On long planetary voyages, operational expenses can easily approach the expense of designing and building the spacecraft divided by 2. Any technologies that reduce expenses associated with flight operations significantly contribute to overall cost reduction.

In a traditional mission, ground controllers track the exact position of spacecraft and receive detailed information about the operation of components. Ground controllers issue commands that adjust the position of spacecraft or cause them to perform special operations. In effect, controllers fly spacecraft from Earth using radio signals and complex digital codes. For planetary missions, ground controllers receive and send signals through the giant dish antennas of NASA's Deep Space Network. Time on the Deep Space Network is precious and expensive.

In the science fiction movie *2001: A Space Odyssey*, astronauts on a journey to the outer solar system depend upon a HAL 9000 computer to monitor the health of their spacecraft. Deep in space, they do not rely upon ground control. The HAL 9000 is an artificial intelligence computer fully capable of operating the spaceship and communicating with the crew in a soft, spoken voice. When the movie first came out, the HAL 9000 was strictly science fiction.

In one respect, science fiction is coming true. Scientists and engineers are building semi-intelligent computers that can fly spacecraft. Such computers require significant excess capacity, the key to spacecraft autonomy. Excess computer capacity allows designers to transfer many of the functions previously performed by ground controllers to spacecraft flying alone. The people who built *Deep Space 1*, the first New Millennium spacecraft, did this in three important ways.

Deep Space 1 contained an AutoNav program that allowed the spacecraft to check and change its location in space. The spacecraft navigated by comparing the position of nearby asteroids to fixed positions of stars. The spacecraft's computer contained the orbits of 250 asteroids relative to the positions of 250,000 stars.

NASA employees, working with specialists at Carnegie Mellon University, installed a "Remote Agent" software program that allowed the spacecraft to manage its own affairs. The program contained an "executive" capable of making decisions, guided by a knowledge of the operational status of spacecraft components, previously programmed constraints, and generalized assignments relayed from the ground. "Compared by some to the HAL 9000 system," designers wrote, "Remote Agent is capable of planning and executing many onboard activities with only general direction from the ground."[7]

Spacecraft with artificial intelligence software on board communicate with ground controllers in different ways. *Deep Space 1* carried a Beacon Monitor Operations Experiment in which the computer reported spacecraft health using four general tones: "green" for acceptable, "orange" for an anomaly resolved by the spacecraft, "yellow" for a need to transmit data, and a "red" tone requiring urgent assistance. Because the tone does not contain detailed digital information, it can be received by a relatively small ground antenna, thus saving the expense of operating a large dish antenna except in special situations.

Mission controllers did not rely upon these devices at all times. Rather, the devices were flown on *Deep Space 1* as experiments to be used during specific test periods. Building a semi-autonomous spacecraft that can control its own operations for long periods of time is more challenging. Designers of the ill-fated *Lewis* satellite installed a Safe Mode feature that allowed their spacecraft to operate without continuous ground control. Like the New Millennium program that produced *Deep Space 1* and *2*, the *Lewis* satellite was part of a Small Satellite Technology Initiative designed to flight-test new instruments and technologies.

Under traditional ground control rules, new spacecraft are monitored continuously during their early flight period. Spacecraft behave in temperamental ways during their first days of flight, and the *Lewis* satellite was no exception. Shortly after launch, its attitude control system faltered and its solid-state data recorder failed. On the third day, ground controllers lost contact with the satellite for two orbits. They found the satellite in an uncontrolled attitude with its batteries partially discharged. Normal operation was restored, and the satellite put into Safe Mode with solar panels pointed toward the sun. Even though ground controllers did not know the cause of the anomaly, they went home.

In a semi-autonomous state, the spacecraft flew itself. Again, minor perturbations in the attitude control system initiated a spin, pulling solar panels away from the sun. Now, however, the spacecraft was on its own. Its processor sensed excessive thruster firings. To correct the problem, the spacecraft disabled its A-side thrusters and switched control to its B-side processor, which sensed excessive firings and shut those down too. Without the thrusters, the spacecraft entered a free drift from which it could not recover without help from below. By the time ground controllers returned to work, the spacecraft had died, its batteries drained.

Members of the failure investigation board did not condemn the philosophy that led the *Lewis* spacecraft designers to experiment with semi-autonomous control.[8] Semi-autonomous operation is a noble goal, and necessary to fly large numbers of low-cost spacecraft. Translating a noble goal into a working technology can be hard. It is full of risk and possible failure, as the *Lewis* satellite experience shows.

ASSESSMENT The final decade of the twentieth century began with the development of the *Cassini* Saturn probe, a spacecraft weighing nearly 3 tons (not counting propellant) and costing nearly $2 billion for design and fabrication. Partly as a challenge, a Mars Pathfinder team working at the same NASA field center as the Cassini group built an 800 pound lander for $175 million and incorporated a 35 pound, high-tech rover at an additional expense of $25 million.

Many people worried that the Pathfinder experience was a "one time only" event, an experiment that could not repeated when scien-

---

6   SELECTED MISSION COSTS (real-year dollars, in millions)

▲ *Cassini.* Launched 1997. Development, $1,422M; launch support, $422M; mission operations and data analysis, $755M; tracking and data support, $54M; foreign contribution, $660M; total, $3,313M.

▲ Mars *Observer.* Launched 1992. Development, $479M; launch vehicle and ground operations, $293M; mission operations and data analysis, $41M through loss of the spacecraft in August 1993; total, $813M.

▲ Mars *Pathfinder.* Launched 1996. Development, $175M; micro-rover, $25M; launch support, $50M; mission operations and data analysis, $14M; tracking and data support, $1M; total, $265M.

▲ Mars *Global Surveyor.* Launched 1996. Development, $131M; launch support, $53M; mission operations and data analysis, $90M; total, $273M.

▲ Mars *Climate Orbiter* and *Polar Lander.* Launched late 1998 and early 1999. Development, $193M; launch support, $92M; mission operations and data analysis, $43M; tracking and data support, $1M; total, $329M.

▲ *Near-Earth Asteroid Rendezvous* (NEAR). Launched 1996. Development, $125M; launch support, $44M; mission operations and data analysis, $42M; tracking and data support, $1M; total, $212M.

▲ *Lunar Prospector.* Launched 1998. Development, $31M; launch vehicle, $26M; mission operations and data analysis, $7M; total, $64.

▲ *Stardust.* Launched 1999. Development, $127M; launch support, $45M; mission operations and data analysis, $37M; total, $210M.

▲ *Deep Space 1.* Launched 1998. Development, $99M; launch support, $45M; mission operations and data analysis, tracking and data support, $11M; total, $156M.

▲ *Deep Space 2.* Launched 1999 with Mars Polar Lander. Development, $25M; launch vehicle integration, $2M; mission operations and data analysis, $1M; total, $27M.

Sources: NASA, "Budget Estimates" (special issues section, various years); NASA, "Mars Pathfinder Project Cost and Funding Summary"; NASA, "Lunar Prospector Press Kit," January 1998.

tists and engineers tried to apply low-cost, low-weight technologies to a broad spectrum of spacecraft. In fact, the next mission replicated the Pathfinder experience. Mars *Global Surveyor*, which reached Mars two months after the *Pathfinder* probe, cost $131 million to develop. Spacecraft weight, at 745 pounds, likewise remained low.

---

7   SELECTED SPACECRAFT WEIGHTS

▲ *Cassini:* orbiter, 4,685 lbs; Huygens probe, 705 lbs; launch vehicle adapter, 298 lbs; propellant, 6,905 lbs; total, 12,593 lbs.

▲ Mars *Pathfinder:* lander, 594 lbs; microrover, 35 lbs; weight at Mars entry, 1,012 lbs; weight at launch, with propellants, 1,760 lbs.

▲ Mars *Global Surveyor:* spacecraft, 1,479 lbs; fuel, 839 lbs; total, 2,318 lbs.

▲ Mars *Climate Orbiter:* spacecraft, 745 lbs; fuel, 642 lbs; total, 1,387 lbs.

▲ Mars *Polar Lander:* lander, 639 lbs; cruise stage, 181 lbs; aeroshell and heat shield, 309 lbs; propellant, 141 lbs; total, 1,270 lbs.

▲ *Near Earth Asteroid Rendezvous* (NEAR): spacecraft, 1,056 lbs; propellants, 704 lbs; total, 1,760 lbs.

▲ *Lunar Prospector:* spacecraft, 348 lbs; propellant, 302 lbs; total, 650 lbs.

▲ *Stardust:* spacecraft, 661 lbs (includes 101 lb return capsule and parachute); propellant, 187 lbs; total, 848 lbs.

▲ *Deep Space 1:* spacecraft, 831 lbs; hydrazine fuel, 68 lbs; xenon fuel, 181 lbs; total, 1080 lbs.

▲ *Deep Space 2:* aftbody (ground station), 3.8 lbs; forebody (penetrator), 1.5 lbs; aeroshell, 2.6 lbs; total, 7.9 lbs.

Utilizing even better technologies, aerospace engineers pushed spacecraft costs even lower. Development expenses fell to $127 million on the *Stardust* probe and $99 million on *Deep Space 1*. Project teams building *Lunar Prospector* and *Deep Space 2* cut development costs again, to around $30 million each. Weights fell as well. *Lunar Prospector*, a spartan vehicle, had one-third of the mass of Mars *Pathfinder*. The design team for *Deep Space 2* produced two probes that weighed only 8 pounds each.

Engineers also tried to suppress operational costs. By spending only $15 million after the spacecraft left Earth, the Mars Pathfinder team set a standard for low-cost mission operations. By contrast, Cassini managers set aside $800 million for spacecraft operations. The Cassini mission was set to fly for nearly 11 years, of course, while Pathfinder took less than one. Still, $15 million for operations and data retrieval is a relatively small sum. This mark proved harder to match. Operational budgets on other "faster, better, cheaper" missions climbed as high as $90 million. On three missions specifically designed to lower operational costs, however, project teams beat the Pathfinder mark: Lunar Prospector, Deep Space 1, and Deep Space 2.

Cost-saving technologies have an accumulating effect when rigorously applied. Smaller spacecraft require less costly launch vehicles; more efficient propulsion and landing technologies encourage smaller spacecraft. The project team for Deep Space 2 approached a threshold of sorts when they developed tiny Martian probes that weighed only 8 pounds each. The ground station for Deep Space 2 was just 4 inches high, topped with a 5 inch antenna. With *Deep Space 2* and the *Sojourner* micro-rover, spacecraft weights fell faster than spacecraft costs. The latter succeeded; the former failed. The history of the micro-electronics and desktop computer industry suggests that when size shrinks, cost reduction is not far behind.

Visionaries anticipate a time when small, low-cost spacecraft penetrate the cosmos. By applying technology in creative ways, that vision may not be as fanciful as it seems. The first few "faster, better, cheaper" projects suggested that standards which seem impossible at their conception can be exceeded by clever scientists and engineers. The standards can even be applied to missions that are by their nature intrinsically large.

RISK AND RELIABILITY

When the "faster, better, cheaper" initiative got started, NASA Administrator Dan Goldin told project workers to take more risks. "Be bold," he said, "push the limits of technology . . . lose a few." (He was speaking about robotic spacecraft.) A space program without risk, he warned, would degrade NASA to a state of "inert uniformity" wherein it would lose the support of Congress and the interest of the American people.[1]

At the same time, everyone wants his or her mission to succeed. After Mars *Polar Lander* disappeared, Goldin felt obliged to reemphasize the importance of reliability. Project teams should be as concerned with risk management as with cost, schedule, and performance, he said, adding a fourth element to his general charge.[2]

In essence, NASA executives told project leaders to take more risks but to ultimately prevail. A few failures along the way were permissible, but in the long run the whole initiative had to produce cost savings, shorter schedules, better performance, and reliable spacecraft. Goldin was particularly upset that some contractors assumed his initiative meant "cheaper" and nothing else. In his mind, cheaper also meant better. Advocates of the initiative wanted leaders of low-cost initiatives to achieve more for each dollar than managers operating in the old, high-cost way.

MANAGING RISK People who build spacecraft face the prospect of failure every day. They make design decisions that must be correct long before their spacecraft even leave the ground. They must anticipate how machinery will operate in the hostile environment of space with methods that imperfectly simulate those conditions.

Spacecraft components interact with each other in ways that no person can fully comprehend. As sociologist Charles Perrow has observed, the greatest dangers in complex technologies arise not from the failure of individual parts, but from unanticipated interactions between them. The most intractable threats to mission success lie in the system as a whole.

Spacecraft managers have developed a retinue of procedures for combating failure, each of which contains its own weakness. Redundancy provides a hedge against subsystem faults, but increases the likelihood of system failures by making spacecraft more complex. Testing imperfectly recreates actual flight conditions and may lead to a false sense of security. Sensors and gauges give false signals. Safety features protect critical components, but complicate mission operations. Systems management draws workers into paperwork studies too far removed from actual hardware where problem recognition is likely to occur.

Given the nature of spacecraft and rocket development, failure is a normal by-product of the task. It is one of the primary ways by which engineers learn how to improve the items they design. Engineers learn through trial and error. In his classic study of engineering mentality, Henry Petroski employs a pun from an old-fashioned phrase ("to err is human"). For the title of his book, Petroski observes that *To Engineer is Human*. Underpinning most great engineering achievements, he argues, is a string of failures. Wernher von Braun and his team of German engineers built reliable rockets because of the meticulous, incremental manner in which they achieved progress by making errors. "You made a mistake; you made it only one time," said one of the original rocketeers. "There is no magic formula for making mechanisms work in all situations," said a member from the Mars Pathfinder team. "The trick is to learn from your (and other people's) mistakes."[3]

The people who run "faster, better, cheaper" projects inherently take risks. They build cheaper spacecraft. They speed up the development process. They do not use all of the traditional tools available for minimizing failure.

From the traditional point of view, such practices should reduce reliability. Traditionally, engineers building cheaper spacecraft with less robust safety features have accepted higher failure rates. They even have a name for it. Cheap spacecraft with less robust safety features are called Class 3 (sometimes Class C) spacecraft. Aerospace engineers expect them to fail more frequently—hence they build and fly more of them to account for the increased likelihood that something will go wrong.

In 1988, Aaron Wildavsky wrote a book entitled *Searching for Safety,* widely cited as a rebuttal to the conventional point of view about dangerous technologies. "The secret of safety," he said, "lies in danger."[4] Citing both technical and biological processes, Wildavsky argued that life becomes safer as people learn how to engage in risky behavior. Rocket flight is full of risk. Exposure to its dangers permits space travel. Similar exposure to other dangers, from heart surgery to business risk, creates the knowledge that allows for advances in medicine and economics. The relationship between risk and technology is not a static one. People can learn from exposure to risk in ways that enrich human life.

Advocates of "faster, better, cheaper" asked project workers to learn more by flirting with danger. Much of what the workers were asked to do was counterintuitive, given the traditional point of view. They were asked to increase reliability on spacecraft with traditionally high failure rates (Class 3 spacecraft) by taking risks with them. They were asked to fight failure without spending more money. They were asked to learn through trial and error without making mission-fatal mistakes.

Given the failures that occurred, this was the most controversial aspect of the "faster, better, cheaper" initiative. Project teams did develop a number of techniques for controlling risk on low-cost spacecraft. At the same time, they lost many of the missions that they flew.

CREATING RISK The air bag landing system developed for Mars *Pathfinder* demonstrates the way in which one team sought to improve reliability by taking more risks. To create a global network of low-cost environmental monitoring stations on Mars, spacecraft engineers needed to drop small packages onto rough terrain without much knowledge about local conditions. That seemed to preclude powered descent, which requires a relatively smooth landing zone. NASA executives pressured Pathfinder managers to develop a different landing approach, a new technology, that would work better, save money, and not weigh too much.

Researchers advanced the idea of enclosing the spacecraft in a giant air bag. This posed a number of technical problems. The air bag had to inflate quickly, in less than a second, just before the spacecraft touched down. It had to be lightweight but strong. Once at rest, the bag had to automatically deflate and be gathered together in such a fashion as to not become entangled in the spacecraft it contained. As a further complication, the air bag system had to be integrated with a heat shield and parachute needed to slow the spacecraft from interplanetary to landing velocities. The spacecraft within the air bag had to automatically turn upright no matter how the air bag came to rest. The whole package had to be outfitted with enough motion sensors and altimeters so that the landing sequence could occur automatically.

Air bags can be quite heavy. The fabric alone on the *Pathfinder* air bag made up 25 percent of the total weight of the landing spacecraft. This precluded traditional safety features, such as redundancy. If project members relied upon redundancy, they would drive up the weight of the landing system to the point that the air bag would no longer cushion the spacecraft's fall. The air bag had to be light, simple, and faultless, working as expected the first time it flew.

Project officials set aside $8.6 million to develop the air bag. This was a tiny sum for a new technology. The entry heat shield, as a point of comparison, cost $8.4 million, and it was a known technology borrowed from the *Viking* landers. Initially, project officials hoped to save money by employing computer technology and nonlinear finite element codes to design the bag. This proved inadequate. With their miniscule budget, project officials fell back on traditional engineering methods. They used intuition, trial, and error.[5]

Project officials contacted ILC Dover for help in designing the bag. ILC Dover is in the fabric business, and manufactures innovative items such as spacesuits and military hardware that inflates. Project workers and corporate engineers took their ideas to NASA's Glenn Research Center Plum Brook facility in Ohio, where they tested them in an enormous 10-story vacuum chamber that could simulate atmospheric pressure on Mars. (Earth pressures would make the bag too stiff for accurate measurements.) The design team built a 50-foot wall, bolted jagged rocks to its surface, and tilted it so as to simulate the angle at which the air bag would strike Mars. To create the expected landing speeds, the research team pulled prototype air bags toward the rocky wall with giant bungee cords.

Team members carefully inflated an air bag and threw it against the wall. Rocks shredded the design. They tried an air bag design that deflated upon hitting the wall. The half-deflated bag bounced and lost more air. When it came down the second time, the lander smashed. Over the next year, the team conducted various tests with many different designs. The wall destroyed nearly all. The design team finally settled on a 17-foot tall airbag, with four chambers and 24 lobes that looked like a molecular model for a chemistry class. The construction that seemed to work best consisted of an inner bladder covered with four additional layers of fabric loose enough to rip away as abrasion occurred. For the fabric, the designers used Vectran, a high-strength fiber used to make sails for high-performance racing craft. To inflate the system, the designers employed small solid-fuel rockets that filled the air bags with sooty exhaust.

Project officials thought that that the air bag would bounce a few times and travel less than 100 yards after landing on Mars.[6] The actual landing proved far more dramatic. The landing site was covered with rocks. The air bag bounced at least 15 times, carrying the spacecraft on a wild ride across the Ares Vallis outflow channel.[7] After the onboard accelerometer sensed that the bags had finally stopped moving, small electric winches pulled cords that opened deflation holes and gathered the fabric into four piles. Three metal pedals enclosing the spacecraft opened in such a way as to automatically place the lander in an upright position. The lander camera came on and revealed that the air bag was stuck on the pedal holding the micro-rover, blocking its path to the ground. Flight controllers raised the pedal, winched in the air bag a bit more, lowered the pedal, and sent the rover on its way.

The air bag system increased risk, but it did not fail. Members of the design team did not exceed their budget and they delivered a flight-ready system that created an alternative to powered descent. They were precluded from using many (but not all) of the traditional techniques for combating failure.

The team that followed the Pathfinder mission to Mars did not adopt the air bag design. In order to save weight, the Mars Polar Lander team relied on a more conventional technology—a soft-landing technique that employed an aerobraking shield, a parachute, and a cluster of descent engines. Ironically, the proven technology failed, while the riskier approach worked. Critics blamed the failure on excessive cost-cutting. The Polar Lander team spent only $15 million on their landing system.[8] This was an astonishingly low figure. The Pathfinder team cut their landing budget to $27 million, a huge reduction in itself.

At the same time, one has to respect the results. The riskier landing system proved more reliable in the two trials. This provided a small measure of credibility to the general notion that safety increases as dangers are better understood.

TECHNIQUES Project teams employing the "faster, better, cheaper" approach used a variety of techniques to maintain reliability without increasing costs excessively.

As a first step toward combating failure, project leaders worked to convince team members that risk was okay. Embracing risk was the first step toward reducing failure. People who accept risk are forced to learn. They become more alert and less complacent. They look for the trouble they know is present, and as a consequence become more conscious of the consequences of their actions. Such conditions create work environments that encourage people to confront risk because everyone accepts that problems are going to occur. Convincing team members to accept an increased level of risk was as important to the success of "faster, better, cheaper" as convincing them to accept cost goals.

Project leaders fought risk with simplicity. A simple spacecraft is easier to understand than a large, complicated one. Fewer people are required to develop it, which enhances the opportunity for teamwork. Simple spacecraft commonly employ single-string systems. This reduces weight, saves money, and avoids complexity. A simple spacecraft might fly with a single computer. The *Lunar Prospector* sent to search for ice on the moon carried no in-flight computer at all. It employed a simple, single-string design of flight hardware guided by a micro-processor that could accept a maximum of 60 commands from a flight control team back on the Earth. The Pathfinder team used "one of everything wherever possible," team leader Brian Muirhead proclaimed, "making an exception only where we couldn't build in enough margin or demonstrate high reliability through testing."[9]

Placing a second computer on a spacecraft, of course, decreases the risk that a malfunction in one will doom the mission. The NASA space shuttle carries five flight computers that guide reentry and contain a program that permits them to vote if they disagree.[10] The absence of such redundancy on "faster, better, cheaper" projects forced workers to adopt compensating schemes.

They engaged in extensive testing. Testing provides assurance that components will perform as planned and provides project workers with the sort of "hands-on" activity that familiarizes them with spacecraft idiosyncrasies. "Test, test, test" became the mantra of Tony Spear, who led the Mars Pathfinder design team.[11] To leave time for testing, Spear insisted that contractors and subsystem managers deliver their hardware no later than halfway through the development cycle. This was a demanding requirement inasmuch as the development phase had already been cut to just three years.

Some teams built extra margin into the systems they installed. "Kill it with margin" is an important solution to the problem of high risk on single-string systems.[12] Margin enlarges the tolerance for miscalculations, prepares the spacecraft for new environments, and protects against unexpected events. To create margin, the *Pathfinder* air bags were thrown during the jagged test wall at a vertical speed of 60 miles per hour, even though the test team knew that that they would probably fall at no more than 30.

Engineers faced with single-string systems also employ techniques such as critical components, cross-strapping, processor reprogrammability, solid-state bulk memory, attitude control sensor overlap, and independent safe holds. Such techniques promote what engineers call "graceful degradation," in which engineers isolate subsystems so that component failures do not cascade into the system as a whole.

Project teams utilized formal risk management studies. Said Spear, "we took risk management very, very seriously. We had monthly assessments of risk and mitigation."[13] Formal risk management is a process by which team members break down systems and assess failure probabilities for each component and how critical it is to mission success. The specific techniques vary from project to project. Some managers use Fault Tree Analysis, in which analysts work backward from undesirable outcomes to the small events that might cause them. Items that are highly critical to mission success or have a high probability of failure receive special attention. Some managers use probabilistic risk assessment techniques, in which analysts multiply component probabilities to calculate the overall chance of mission success.

Following risk management reviews, team members are supposed to prepare mitigation plans for critical items. This typically leads to more testing to identify the exact points at which subcomponents might fail and more testing to assess prospective solutions. Testing might reveal, for example, that a particular component fails to operate 10 percent of the time under conditions that are likely to occur 10 percent of the time, producing an overall failure probability of 1 percent. Twenty such components, if they are critical to mission success, will reduce the chances of mission success to just 81 percent (0.99 to the 20th power).

Mitigation plans were submitted for peer review. Within this process, groups of peers from outside the project organization review the precise methods by which design teams calculate probabilities. The peers inspect mitigation plans. They make suggestions, which the design team can accept or reject. Ultimately, responsibility for reducing failure on risky components falls back on the design team. "We never stopped thinking about the risks and making it better," said Brian Muirhead from the Mars Pathfinder team. "Everybody at the engineering level and at the systems level owned the result and really worked hard."[14]

Project leaders confronted with risky components tried to balance them with proven technologies. The air bag system on Mars *Pathfinder* was a risky technology. It worked in conjunction with a heat shield, parachute, and cluster of solid rockets. The parachute had to work perfectly, slowing the spacecraft from a velocity of 900 miles per hour to just one-sixth that speed, and it had to weigh no more than 10 kilograms. Testing the parachute was difficult and prohibitively expensive. The most reliable tests capable of replicating the thin Martian atmosphere required team members to drop the parachute from a high-altitude airplane. Team members were willing to conduct low-altitude tests that simulated the final stages of descent, but they did not want to spend project funds on expensive high-flying tests.

A newly designed parachute would have added even more risk to a system that had already become very risky. To reduce that risk, and save money, the Pathfinder team borrowed the parachute design that engineers from the Viking descent group had used to slow their landers some 20 years earlier. They trusted that the results from the *Viking* high-altitude drop tests would be applicable to *Pathfinder* if they used the same equipment. Integrating a proven technology into a high-risk technology saved money and reduced the likelihood that the entire system would fail.

None of these techniques, individually or in combination, are foolproof. One year from the fixed launch date, members of the Pathfinder team took their solid rocket braking subsystem to the China Lake Naval Weapons Center in the California desert to see how it would perform. The solid rockets had to fire in such a way as to bring the *Pathfinder* spacecraft to a dead stop, still hanging from its parachute, some 40 feet above the Martian soil. The initial test results looked fine, but closer inspection revealed an anomaly in the thrust profile that jeopardized the ability of rockets to stop the spacecraft's descent at exactly the right time.

The anomaly defied explanation. Team members could not identify its cause. If they conducted formal studies, members knew that they would miss their launch date. If they ignored the anomaly, it could doom the mission. In preparing for the test, engineers had removed nearly all of the aluminum from the rocket fuel. Aluminum helps to stabilize rocket burns. Test engineers had removed the aluminum so as to avoid contaminating the structure holding the rockets. Intuition pointed to the absence of aluminum as the likely cause. If intuition was wrong, the landing system would not be ready in time and the mission would fail.

Project leaders told the engineers to put the aluminum back in the rocket fuel, rebuild the engines, and test them in a backshell nearly identical to the one that would fly to Mars. "The new engines were built and the final test completed just six months before launch. The test was completely successful."[15]

IS FAILURE INEVITABLE? Struggles over risk on "faster, better, cheaper" projects are part of a larger debate taking place about high-risk systems in general. That debate concerns the degree to which people can develop procedures that promote reliability in risky technologies.

On one side of the debate are the pessimists, who believe that all efforts to prevent failures in highly technical systems are ultimately self-defeating. They argue that complex systems such as nuclear power generating plants and space shuttles possess features qualitatively different from older technologies such as automobiles which make the former highly susceptible to failure. Organizational analysts like Charles Perrow insist that catastrophic accidents in such systems are "normal" given the way in which these systems are designed.

On the other side of the debate are the optimists, who believe that humans can create complex systems like the Internet that operate at high levels of reliability. The features of "close coupling" (very little warning time) and "complex interactions" need not create failure, they say, providing the operation is organized in the appropriate way.

Both sides agree that modern systems like spacecraft contain features that make them especially susceptible to catastrophic accidents. Both agree that conventional bureaucracies, if employed to manage those technologies, make matters worse. Bureaucracy is a form of organization effective for providing routine, continuous services that do not rapidly change. In the fast-moving realm of high technology, it can produce disastrous results.

During the 1980s, a multidisciplinary group of scientists at the University of California at Berkeley investigated a set of organizations handling hazardous technologies that nonetheless had attained very high levels of reliability. Todd LaPorte, a political scientist and one of the leaders of the group, objected to the "optimistic" label that others gave them. "We (the Berkeley High Reliability Organization project members) are more pessimistic than the other guys. We've seen what it takes to effect conditions of reliability. . . . It's damn demanding, is very costly, and takes remarkable dedication."[16]

Spaceflight, like other high-technology endeavors, is an inherently risky undertaking. People who have studied spaceflight disagree on whether it can be conducted reliably. Some think that reliability will improve; others think that catastrophes will always occur.[17]

People who advocate "faster, better, cheaper" side with the optimists. They believe that spacecraft managers can take high risks and succeed. They carry one additional belief, however, that distinguishes them from high-reliability advocates like LaPorte. They believe that it can be done cheaply.

Studies show what aerospace engineers know intuitively: cost and schedule for a given level of reliability increase with spacecraft complexity.[18] Complex spacecraft are by their very nature more costly. Being more susceptible to interactive failures, they are also less reliable. From this perspective, the route to reduced cost and increased reliability is clear. Build smaller, less complex spacecraft.

To a certain extent advocates of the new initiative did this. They built smaller and lighter spacecraft with lower operating requirements. Launch mass is a major indicator of spacecraft complexity (see box 8), so reductions in size indirectly decrease cost. Being smaller, they require less electric power, another factor by which complexity is gauged. They contain fewer redundant subsystems. This helps to reduce size and some of the propensity for interactive failures.

Relative to their larger counterparts, "faster, better, cheaper" spacecraft are less complex. Compared to traditional Class 3 satellites, however, they are more complex. Smaller does not always guarantee simplicity. Pound for pound, the newer spacecraft tend to be "denser" than their traditional counterparts.[19] Density means that project workers pack lots of high-tech components onto a smaller

---

8 FACTORS CONTRIBUTING TO SPACECRAFT COMPLEXITY

- Launch mass (kg)
- Design life (months)
- Maximum distance from Earth orbit (astronomical units)
- Beginning of life power (watts)
- End of life power (watts)
- Solar array area (m²)
- Solar cell type (Si versus GaAs)
- Array/antenna configuration (body-fixed, deployed, articulated)
- Battery type (lead-acid, NiCd, SniCd, NiH2)
- Battery capacity (A-hr)
- Structures material (aluminum, composite)
- Attitude control type (none, grav-grad/spin, 3-axis)

- Number of p/l instruments
- Pointing accuracy (degrees)
- Pointing knowledge (degrees)
- Number of thrusters
- Propulsion type (none/cold gas, monopropellant, bipropellant/ion)
- Downlink communications band (UHF/VHF, S-band, L/X-band)
- Maximum downlink data rate (Kbps)
- Solid-state recorder memory (mbytes)
- Thermal control type (passive, semi-active, active)

Source: David A. Bearden, "A Complexity-Based Risk Assessment of Low-Cost Planetary Missions," Aerospace Corporation, 2000.

---

frame. As part of their complexity scale, Sarsfield and Bearden include factors such as solar array material, battery type, attitude control system, and fuel type. On those criteria, "faster, better, cheaper" spacecraft rate high. Not unexpectedly, they cost more per pound than traditional, small satellites.

The balance between size reduction and density creates spacecraft for which the "zone of reliability" is quite small. For such spacecraft, the upper edge of the zone is bounded by cost and schedule goals ("faster, cheaper"). The lower edge of the zone is created by complexity. Any attempt to reduce cost or schedule below the appropriate complexity level invites failure; any attempt to buffer failure with excessive spending (or time) violates the premise of the initiative. A narrow range exists between the two.

"Faster, Better, Cheaper"

▲ Risk is inevitable in complex, tightly coupled systems. It is often the source of creativity and innovation.

▲ Reliability, speed, and low cost can be achieved simultaneously by reducing the scale and complexity of projects. Risk need not necessarily lead to failure.

▲ Redundancy and safety features are useful methods for increasing reliability, but they are ultimately self-defeating because they increase cost and complexity.

▲ Self-learning organizations successfully combine a number of apparent contradictions: centralization and decentralization; flexibility and inflexibility; error avoidance and learning through trial and error.

▲ The most important techniques for achieving high levels of reliability involve the creation of self-managing teams.

▲ Formal system reviews are useful, but not the primary means for promoting reliability. The primary means are techniques such as co-location, multitasking, seamless management, shared objectives, extensive testing, hands-on activity, and other approaches designed to create a capacity for problem solving and personal responsibility. Risk management studies and peer reviews are important components in that approach.

▲ To foster a culture of reliability, project teams must be given a high degree of technical discretion.

Traditional Systems Management

▲ Risk can be managed through good organizational design and practices.

▲ Risk, schedule, and cost are interdependent. It is impossible to reduce one without increasing the others.

▲ Redundancy and safety features are primary means for reducing the likelihood of interactive failures.

▲ The combination of centralized systems management with decentralized technical capability produces the tension that contributes to project success.

▲ The most important technique for achieving high levels or reliability is the establishment of formal systems management within a strong project organization.

▲ Formal reviews and paperwork studies are the primary means for controlling cost, schedule, and reliability when dealing with complex technologies.

▲ Technical discretion is important, but it is ultimately incompatible with the need for project oversight and fiscal accountability.

The establishment of such a zone of reliability assumes that the management methods employed to conduct the mission are appropriate to its scale and complexity. There are not many margins for error. With too few resources (time and money), managers lose the margins they need to test and review. If the management system is weak, they lose capacity. In that case, the zone of reliability shifts to the right, into an area where the resources needed to achieve reliability exceed the provided amounts.

People who study high-reliability organizations are correct in one major respect. Managers have created organizations capable of producing nearly error-free activities. Within NASA, they have also done this at significantly reduced costs.

The skeptics are correct in another regard. It is awfully hard to hold projects within the zone of reliability. Too little time, too little money, or inadequate management methods cause projects to drift out of their zones of reliability. This accounts for the abnormally high failure rates experienced by managers conducting "faster, better, cheaper" projects. The concept is workable, but hard to maintain. It requires a vigilance not typically encountered in everyday life.

FUTURE IMPLICATIONS

The "faster, better, cheaper" story sits within a larger history of cost reduction in high-technology undertakings. Modern economies driven by technology derive their power from the ability of entrepreneurs to reduce cost while improving performance. The conquest of space is part of the high-technology revolution. Logically, it would seem, people who build spacecraft should be able to achieve similar gains.

Advocates of "faster, better, cheaper" followed a well-tested route toward this goal. They utilized lightweight technologies to reduce cost and complexity. Through a variety of management reforms, they downsized the teams necessary to carry out this work. They attempted to create a culture in which people believed that cost reduction was as important as mission reliability and they adopted techniques designed to maintain reliability without significantly increasing costs.

The advocates did so against a 35-year history of resistance to cost reduction in the aerospace field. Unlike the history of aviation and micro-electronics, the cost of spaceflight did not fall as the endeavor matured. The cost of conventionally designed satellites and planetary spacecraft hit a flat plateau. The cost of rocketry grew with time.

"Faster, better, cheaper" advocates sought to change this situation. They experimented with alternative methods for building and flying robotic spacecraft. In the process, they identified cost-saving techniques that managers applied to activities of ever-increasing size. The effort to apply this philosophy extends even to the most complex undertaking of all—a human expedition to Mars.

THE COST OF SPACEFLIGHT In the area of space exploration, costs have remained persistently high. Forty years of spaceflight have not produced the same sort of productivity gains that entrepreneurs in other high-technology industries have achieved.

The history of planetary spacecraft demonstrates this reality. Between 1962 and 1973, the United States launched 10 *Mariner*-class spacecraft to explore the inner solar system. The first set of spacecraft were relatively primitive and expensive to build. *Mariners 1* and *2* were designed to fly to the vicinity of Venus; *Mariners 3* and *4* were aimed at Mars. The four spacecraft each weighed about 500 pounds. In the value of year 2000 dollars, the price of the four spacecraft totaled nearly $1 billion. That produces an overall cost of about $1 million per pound (inflation-adjusted dollars). Development cost relative to spacecraft mass provides a rough measure for comparing efforts at cost control. (For spacecraft developed in pairs, only the weight of the first spacecraft is counted, since the incremental cost of fabricating a second model tends to be relatively small. Launch vehicles and operational costs are not included in this analysis.)

As the Mariner program progressed, expenses declined. In constant dollars, the cost of producing *Mariners 6* and *7* was about three-fourths of the expense of *Mariners 1* through *4*. *Mariners 8* and *9* cut that further. Concurrently, spacecraft engineers learned how to pack on more equipment for less money. Spacecraft weights grew to 838 and then 2,196 pounds and capability improved. Unit costs fell to about $300,000 per pound.

Having reached the $300,000 level, the unit cost of planetary spacecraft fell no more. Spacecraft weights, nonetheless, continued to grow. The *Galileo* probe to Jupiter weighed 4,500 pounds and the *Cassini* probe to Saturn weighed 5,500 pounds. Viking program managers produced spacecraft that topped the scales at 7,700 pounds for each orbiter-lander pair. With cost per pound relatively fixed,

10 SELECTED SPACECRAFT DEVELOPMENT COSTS

▲ *Mariners 1, 2, 3, 4.* Flybys of Venus and Mars. Launched 1962 and 1964. Spacecraft development cost: $141 million ($955 million in 1999$). Produced two pair of spacecraft each weighing 446 and 574 pounds. Cost per pound: $936,000 (1999$).

▲ *Mariners 6 & 7.* Flybys of Mars. Launched 1969. Spacecraft development cost: $127 million ($711 million in 1999$). Produced one pair of spacecraft each weighing 838 pounds. Cost per pound: $848,000 (1999$).

▲ *Mariners 8 & 9.* Orbited Mars. Launched 1971. Spacecraft development cost: $135 million ($667 million in 1999$). Produced one pair of spacecraft each weighing 2,196 pounds. Cost per pound: $304,000 (1999$).

▲ *Viking.* Two orbiters and landers to Mars. Launched 1975. Spacecraft development cost: $875 million ($3.7 billion in 2000$). Produced two pair of spacecraft each weighing 7,700 pounds. Cost per pound: $480,000 (2000$).

▲ *Cassini.* Orbit Saturn and drop Huygens probe on Titan. Launched 1997. Spacecraft development cost: $1,375.9 million ($1,650 million in 1999$). Produced spacecraft and probe weighing 5,551 pounds. Cost per pound: $297,000 (1999$).

▲ Mars *Observer.* Orbit Mars. Launched in 1992. Spacecraft development cost: $479 million ($663 million in 2000$). Produced one spacecraft weighing 2,240 pounds. Cost per pound: $296,000 (2000$).

▲ Mars *Global Surveyor.* Orbit Mars. Launched in 1996. Spacecraft development cost: $131 million ($145 million in 2000$). Produced one spacecraft weighing 1,479 pounds. Cost per pound: $98,000 (2000$).

▲ Mars *Pathfinder.* Landed on Mars in 1997. Spacecraft development cost: $195 million, including micro-rover ($220 million in 2000$). Produced one spacecraft and micro-rover weighting 1,256 pounds (excluding cruise stage). Cost per pound: $175,000 (2000$).

▲ *Chandra X-Ray Observatory.* Space telescope. Launched 1999. Observatory development cost: $1,581 million ($1,818 million in 1999$). Produced one telescope weighing 10,560 pounds (excluding propellants and pressurants). Cost per pound: $172,000 (1999$).

▲ *Space Infrared Telescope Facility* (SIRTF). Space telescope. Launch planned for 2002. Observatory development cost: $473 million ($473 million in 2000$). Produced one telescope weighing 1,650 pounds (including 250 liters of liquid helium). Cost per pound: $287,000 (2000$).

and spacecraft growing in size, mission costs increased rapidly. A record of sorts was created when the combination of high weight and fixed costs produced missions where spacecraft development alone surpassed $3 billion in inflation-adjusted dollars.

The forces that caused unit costs to fall in other high-technology sectors did not affect the realm of spaceflight. Corporate leaders count on competition to enforce a culture of innovation and cost reduction. Cold War competition with the Soviet Union produced governmental generosity, however, not parsimony. Private entrepreneurs know that free markets impose pricing discipline. For the first 40 years, most space activities were supported by tax dollars and directed by government bureaucracies insulated from the pressures of open markets. Industrial executives count on learning curves and mass production to reduce per unit cost. Spaceflight engineers developed a culture in which people constantly sought to produce something new, precluding what cost advantages might accrue from accumulated wisdom. Business executives seek to simultaneously improve cost, schedule, performance, and reliability. People promoting space exploration created a culture in which workers assumed that schedule, performance, and reliability had to be purchased with greater spending.

In the final decade of the twentieth century, the "faster, better, cheaper" initiative appeared. Its advocates made two fundamental changes. Using micro-technologies and design simplification, they built smaller, lightweight spacecraft. Using team-based management techniques, they reduced the size of their workforces.

As a consequence of such efforts, aerospace engineers produced spacecraft weighing, in the case of Mars *Pathfinder* and *Global Surveyor*, about 1,300 pounds. Consequently, they pared spacecraft development costs below the $300,000 per pound plateau. Together, light weight and low cost produced spacecraft with development costs less than $200 million. Total program cost as a consequence fell to a fraction of that incurred associated with *Cassini*- and *Viking*-class missions.

EXTENDING THE APPROACH As the "faster, better, cheaper" approach gained credibility, aerospace executives began to apply it to projects both medium and large. The concept was applied to the *Space Infrared Telescope Facility* (SIRTF), to the effort to create cheap access to space (CATS), and to project planning for a human expedition to Mars.

In effect, this represented an effort to apply elements of the approach to Class 1 and Class 2 spacecraft. By definition, Class 1 and Class 2 spacecraft are high-visibility missions that cannot be allowed to fail. To skeptics who viewed the initiative as nothing more than an effort to reproduce Class 3 spacecraft, this seemed particularly dangerous. By definition, Class 3 spacecraft are moderate risk missions with higher propensities to fail. Any effort to fly a Class 3 spacecraft on a Class 1 mission, in the minds of skeptics, invites catastrophe.

The strongest advocates of the approach did not believe they were producing Class 3 spacecraft. They believed that they had developed methods of cutting costs while maintaining reliability. That belief lent confidence to the efforts to apply the techniques to larger and more vulnerable missions.

In all of the larger missions, advocates of the approach sought to cut spacecraft weight through changes in technology. One of the best examples was the *Space Infrared Telescope Facility* (SIRTF).

Space telescopes can be terribly expensive affairs. In the dollars of its day, the *Hubble Space Telescope* cost $2 billion to design and build, not counting the cost of launching it into orbit or operating it once there. More recently, the *Chandra X-Ray Observatory* ran up a $1.6 billion development bill. When asked to estimate the price of an orbiting infrared telescope, experts predicted that it would cost $2.2 billion. That only encompassed design, test, and fabrication. It did not include launch expense, possible cost overruns, and the substantial funds needed to operate the observatory once in space.

Astronomers have wanted to build large space-based infrared observatories for as long as they have wanted to put telescopes in space. Infrared telescopes measure the heat given off by distant objects. They can detect newborn stars, extra-solar planets, and objects whose visible light does not pierce the cosmic dust curtains that obscure so much of the universe. They can study galaxies and quasars formed at the beginning of time, objects whose emissions have shifted into the infrared portion of the spectrum. They can do all these things, but not from the Earth, where ambient temperatures and heat omissions obscure the view.

NASA officials began studying the concept in 1971; they hoped to launch such a telescope in 1987. A space-based infrared telescope promised to complement the work performed by the other great observatories: the *Hubble Space Telescope*, the *Compton Gamma Ray Observatory*, and the *Chandra X-Ray Observatory*. At the time, NASA officials hoped to build a 22-foot-high infrared observatory weighing over 12,500 pounds, including the 3,800 liters of liquid helium needed to cool the instrument, and launch it in the space shuttle. After years of study, lawmakers told NASA to forget the infrared observatory. At more than $2 billion, it was simply too expensive to build.

Rather than lose the observatory, space scientists redesigned the facility so that it weighed only 1,650 pounds. In the process, the telescope lost little of its capability. Lower weight allowed it to fit on top of a much less expensive *Delta* 7920-H launch vehicle. The cost per pound of the redesigned observatory remained in the $300,000 range. The new design weighed so much less than its predecessor, however, that the overall expense of developing the telescope fell to $473 million.

Project planners cut weight through a combination of technical changes that they characterized as thinking "outside the box."[1] First, they changed the position of the telescope. Instead of having it circle the Earth, as originally planned, they fashioned an orbit in which the telescope would follow the Earth around the sun, trailing farther and farther behind. An Earth-trailing heliocentric orbit creates a benign thermal environment, since the telescope is not constantly adjusting itself to the changing position of heat sources like the sun and moon.

Adjustments in heat constraints allowed project leaders to launch the telescope "warm," relying upon radiative cooling to reduce instrument temperature before beginning to consume cryogenic coolant. This significantly reduced the amount of liquid helium that the spacecraft had to carry.

Advances in infrared array technology, prompted by the military's need to detect heat sources like missile launches, ensured high-quality observations. Advances in solid-state memory allowed project planners to adopt a "store and dump" approach that cut data transmission costs. Mission planners did make one compromise in capability. They cut the planned life of the facility from the original goal of 5 to 2.5 years, but added enough coolant so that the instrument might continue to function for the original period of time. As a result of changes such as these, project planners were able to design a much smaller instrument. It contained just 16 percent of the mass of the third great telescope, the *Chandra X-Ray Observatory*, and cost much less.

CHEAP ACCESS TO SPACE Rocket scientists want to cut the cost of moving payloads from Earth into space. Cheap access to space is easy to imagine and monstrously difficult to achieve. Reducing the cost of orbital access has proved to be one of the most intractable problems of spaceflight.

Spaceflight specialists measure transportation costs by comparing the cost of delivering a given payload mass. The *Saturn V* rocket that lifted Americans to the moon delivered its 285,000-pound payload to low Earth orbit at a cost of about $4,500 per pound, in inflation-adjusted year 2000 dollars. The expense of a *Titan III C*, which could lift 23,000 pounds during the same era, was about $4,700 per pound (also inflation-adjusted). Aerospace executives predicted that a reusable space shuttle would cut those costs by a factor of 10. In fact, the expense almost doubled. The cost of using the space shuttle to launch the *Chandra X-Ray Observatory* worked out to about $8,000 per pound. That is an exorbitant amount, well beyond the goals originally set for space transportation.

The shuttle experience proved that cost per payload pound could not be reduced simply by employing reusability. The shuttle is reusable but such a tender technology that the orbiter spacecraft must be disassembled and rebuilt after every mission. The shuttle is difficult to launch and costly to fly.

As the twenty-first century dawned, people in a number of organizations struggled to design more efficient launch vehicles. One of the most creative attempts was made by a small group of private entrepreneurs gathered in a corporation called Rotary Rocket. They sought to alter launch technology in a number of ingenious ways. They tried to eliminate the heavy and expensive turbopumps needed to deliver fuel to traditional rocket engines, substituting instead a whirling ring that used centrifugal force to spin fuel and oxidizer to 72 small combustors. For rocket components, they turned to composite shells in the hope that such components could be mass-produced from inexpensive molds. For landing, engineers placed a beanie-like rotor on top of the rocket. During takeoff, the helicopter-like blades folded into the spacecraft. During descent, they extended, slowing the reusable craft as it fell.

Their rocket design was extraordinarily light. On paper, the *Roton* rocket weighed 22,000 pounds. The whole space shuttle, by contrast, weighs 625,000 pounds in its inert state, without payload or fuel. (That includes the external fuel tank and solid rocket boosters.) Fully fueled, the *Roton* rocket was designed to deliver 7,000 pounds to low Earth orbit—a remarkable 3-to-1 ratio of spacecraft mass to payload. The space shuttle, by contrast, can deliver 55,000 pounds—a much higher and less efficient ratio of 11–1. The *Roton* rocket design may not succeed, but the underlying philosophy represents the future of launch design: use technology to deliver more payload with less weight.

Not every payload fits on a small rocket like the *Rotary* design.

The conquest of space also requires rockets that can launch 45,000 pound payloads and more—and do so at substantially lower costs. NASA Administrator Daniel Goldin summarized the challenge in the following way: "We want to develop technologies that will allow industry to build a vehicle that takes days, not months, to turn around; dozens, not thousands of people to operate; reliability ten times better than anything flying today; and launch costs that are a tenth of what they are now."[2]

During Goldin's tenure, NASA invested heavily in the X-33 test vehicle program. Shaped like a lifting body, the *X-33* was not designed to fly into orbit. Rather, it supported the development of technologies necessary to build larger vehicles that could. NASA gave the contract to develop the *X-33* to the Lockheed Martin company in the hope that Lockheed Martin would use it as a model to build a much-larger *VentureStar* vehicle.

On paper, *VentureStar* was very impressive. Design plans called for a single-stage, reusable spaceship capable of delivering shuttle-size payloads. It would dart into orbit without shedding fuel tanks or booster rockets, then fly back to Earth and land like an airplane. As initially planned, it would do this automatically, without pilots onboard. People, if they wanted to ride on *VentureStar*, would sit in a standard-size container fit into the cargo bay. Early design plans called for a spacecraft weighing only 212,000 pounds—about one-third of the space shuttle mass.

To create a lightweight launch vehicle, project planners employed special technologies both new and old. They planned to install a thermal protection system for dissipating reentry heat built out of a thin titanium alloy developed for the X-15 test program 40 years earlier. They tried to build lightweight fuel tanks out of a graphite-epoxy composite, baked in huge ovens. This proved quite difficult. "X stands for experimental," warned an official from the company baking the tanks. During early work on the *X-33*, the molding process failed. "These things happen," the official observed.[3]

The most innovative technology on the *X-33* was its aerospike engine. The aerospike engine eliminates the bell-shaped nozzles common to early rocket design. An aerospike engine is more compact and integral to the airframe, reducing both weight and drag. Shaped like a *V* (which symbolizes the "spike"), the engine consists of a series of small combustion chambers mounted on the top edge of a curving frame. The combustion chambers are electronically controlled, thereby eliminating the heavy mechanical gimbals needed to tilt conventional engines. The aerospike design has a strange airflow characteristic that improves its performance as it ascends through the thinning atmosphere. At high altitudes, gases expand and add thrust by pushing against the *V*.

The overall effort to develop a low-cost launcher that could reach orbit in a single stage proved challenging. In 1999, after failing to raise the money necessary to build the first flight version of the *Roton*, executives at Rotary Rocket reorganized the firm, laid off workers, and abandoned their technically challenging engine design. Engineering problems plagued the development process for the *X-33*, and costs increased in a manner normally not tolerated on "faster, better, cheaper" projects. Industry observers doubted whether private investors would finance *VentureStar*, given technical difficulties and a paltry launch market, and NASA executives cancelled funding. According to one congressional overseer, the development effort was "one of the toughest things we've ever done in space."[4]

In the 1970s NASA officials set similar goals for the space shuttle, which did not meet its cost objectives. In the 1980s engineers failed to build a National Aerospace Plane. Experts worried that launch complexity would overpower the desire for low-cost transportation. As of 2000 no one had produced a low-cost launcher, but advocates of cheap spaceflight continued to work toward this dream.

MISSION TO MARS The most ambitious application of the "faster, better, cheaper" approach occurred around planning efforts for a human expedition to Mars. A Mars mission would be a undertaking of unimaginable complexity. Humans would leave the Earth and spend 3 years in a radiation-soaked vacuum. Mission planning would be exhausting, the challenges of coordination perplexing. Yet advocates of the approach believe that the initial mission could be accomplished for about one-third of the cost (in inflation-adjusted dollars) of the first expedition to the surface of the moon.

The Apollo program cost $21.3 billion through the first landing on the moon—the equivalent of about $120 billion in year 2000 dollars. First estimates for President George Bush's Space Exploration Initiative, which included a mission to Mars, topped $400 billion—the equivalent of $550 billion in 2000. "Once somebody tagged $400 billion on it," observed a senior NASA official, "it was dead on arrival."[5]

Following the demise of the Space Exploration Initiative, NASA Administrator Daniel Goldin challenged planners at the Johnson Space Center to develop a very-low-cost scenario that could be accomplished in about eight years. Space advocates hoped to pitch the mission to an incoming president with the promise that the first crew would reach Mars before the end of the president's second term. "It has a fundamental appeal," observed Robert Zubrin, founder of the Mars Society. A president concerned with his or her legacy would be hard pressed to refuse it, "unless the guy is made out of wood."[6]

Estimates for a low-cost mission to Mars varied from $6 to $50 billion. Robert Zubrin, one of the leading architects of the low-cost scenario, believes that NASA could send humans to Mars for $25 billion-an appropriation of about $3 billion annually during the eight to 10 years needed to prepare the voyage. Without government interference, Zubrin suggests that private entrepreneurs taking greater risks could dispatch the first expedition for about $6 billion. Planners at NASA's Johnson Space Center used a $50 billion target.[7]

A low-cost Mars expedition would have to be a lightweight mission. Lightweight technologies would be needed to lower the cost of launching materials from Earth and lower the weights of materials

11 COST OF PROJECT APOLLO THROUGH THE FIRST MISSION TO THE SURFACE OF THE MOON (real-year dollars, in millions)

Apollo spacecraft.	$6,939
Saturn launch vehicles & engine development.	8,794
Operations, tracking, & data acquisition.	1,678
Facilities.	1,810
Operation of NASA Centers.	2,128
Total.	$21,349

Source: T. O. Paine letter to Clinton Anderson, November 21, 1969, with attachment.

needed on Mars. The most dramatic suggestion for lowering weight was made by Zubrin in a book entitled *The Case for Mars.* Zubrin suggested that the first human explorers on Mars "live off the land," much as terrestrial explorers like Lewis and Clark had done while transgressing the Earth. Specifically, Zubrin suggested that project engineers send factories to Mars capable of manufacturing rocket fuel and other consumables such as water and oxygen. In theory, all such elements could be extracted from Martian ice or from the local atmosphere.

On paper, the scenario looks impressive. But are the cost goals feasible? A low-cost Mars mission is based on two assumptions, both derived from spaceflight activities underway as the twenty-first century began. Rocket scientists must reduce the cost of lifting payloads to Earth orbit to the much-desired $1,000 per pound and aeronautical engineers must design interplanetary spacecraft that cost little more than the International Space Station.

Mission planners propose to build a Magnum Launch Vehicle costing no more than $5 billion to develop that can fly payloads to Earth orbit for no more than $1,000 per pound as operations proceed. This would allow expedition leaders to deliver the required payload-about 800,000 pounds-to low Earth orbit at a total cost not to exceed $6 billion. Planners believe that this can be done by judiciously using shuttle launch facilities and deriving liquid fly-back boosters from the current shuttle configuration.

To go to Mars, humans would need to push six spacecraft each weighing about 135,000 pounds from low Earth orbit to Mars. Mission plans call for two cargo vehicles that would land on Mars, two transit vehicles that would remain in orbit around Mars and provide the means for the return voyage, and two habitats that would also reach the landing zone-the last one containing the crew. Planners hope to restrain the expense of these spacecraft by deriving their design from components built for the International Space Station. The International Space Station is a test bed for interplanetary flight technologies, one of the silent objectives of the program. It is silent in the sense that government officials cannot openly discuss it because Congress has not approved any flights to Mars. Any Mars habitats and transit vehicles will be based extensively on space station design, with the obvious difference of engines attached. Habitat and transit vehicle designs will be tested by attaching models to the station superstructure.

The International Space Station weighs slightly more than 1,000,000 pounds. It will cost the United States about $26 billion to develop, plus about $15 billion to transport to orbit. (International partners add their own funds and may be expected to contribute to a Mars expedition as well.) By holding the cost of spacecraft development below $26 billion, Mars planners have a fighting chance of meeting their cost goals. They could spend $6 billion for transportation, less than $26 billion for spacecraft development, and have funds left over for operations and additional engine development.

A mission to Mars, however, is considerably more complex than an orbiting space station. The International Space Station orbits in a relatively benign environment, neither landing not ascending from another sphere. If mission planners were to use the Apollo Moon project as the analog for a Mars expedition, projected expenses would quickly soar. Compared to a flight to the moon, a Mars mission would need eight times as much equipment, be gone 225 times as long, and travel 600 times as far. If humans used Apollo-type methods to go to Mars, given the added level of complexity, expenses would climb well above $1 trillion.

Mission planners, of course, do not want to use Apollo-type methods. They want to substitute Pathfinder methods for Apollo-type technology. The Pathfinder team accomplished its mission for $265 million. For that, they launched a canister weighing 1,760 pounds and sent it to Mars. A human expedition to Mars would require 460 times as much payload on a mission seven times as long and the payload would have to come home. Added mass, mission duration, and other factors increase complexity. Added complexity (plus the propensity of costs to increase exponentially with it) escalate mission cost. A conservative interpretation of mission complexity extrapolated on the basis of the Pathfinder experience yields total mission costs in excess of $200 billion. The ability of planners to confine costs to the $6 to $50 billion range requires methods more radical than those used on *Pathfinder*.

Mission complexity and payload weight drive costs up; simplicity and technological maturity drive costs down. With time, the cost of a Mars expedition may fall. Humans will gather experience with long-duration spaceflight through the International Space Station. Through test and development work, they will improve launch capabilities. Engineers will test Mars habitat modules and other technologies by attaching them to the space station or landing them on the Moon. They will experiment with in situ propulsion manufacturing plants on Mars, using robotic missions. They will conduct additional low-cost missions, including robots that will explore and return samples from Mars. All such investments help to lower the cost of interplanetary exploration below the $200 billion range.

The first step-reducing the cost of a Mars expedition from Apollo- to Pathfinder-type levels-is by no means assured. The Pathfinder approach has never been applied to a large human space flight endeavor. It would require radical changes in mission management and technology. Most important, it will require a vastly simplified management structure, one that lowers mission costs by creating a small, cohesive project team. Many people believe that a Mars mission is too large to be managed by an undersized team.

LESS WEIGHT, LEAN TEAMS Building an effective project team that can restrain costs while achieving reliability is just as hard as rocket science. To achieve this goal, managers overseeing a number of advanced projects have simplified their development organizations. The Rotary Rocket team used a very lean management system. The original concept was developed by a skunk works group of about 50 employees located in southern California, and assisted by about 200 contractor employees at sites around the United States. In the tradition of aviation pioneers, the Rotary Rocket team also sought the advantages of a private firm unencumbered by the restrictions of government bureaucracy. The managerial advantages of entrepreneurship, however, were muted by the financial disadvantages of having to raise private funds. One space analyst worried that little companies like Rotary Rocket "just don't have enough zeros in their budget to succeed."[8] Rotary executives won attention for creativity, although they experienced difficulty finding the funds necessary to build machinery they could fly.

Lockheed Martin executives set up a skunk works team to develop the *X-33*. As with Rotary Rocket, development work was carried out by private industry, maximizing the technical discretion of project engineers and freeing them as much as possible from bureaucratic and political oversight. Private management, however, implied private funding. In exchange for the tax-funded *X-33* contract, industry executives pledged to develop *VentureStar* using venture capital, but lost enthusiasm for this approach once tax funding began to disappear.

People overseeing the *Space Infrared Telescope Facility* adopted management techniques that strengthened teamwork. NASA executives relocated the project to the Jet Propulsion Laboratory, where experience with "faster, better, cheaper" techniques was strong. Executives limited the development phase to four years and team leaders invited contractors onto the core group early during the design phase so as to promote cooperation. Planners for the Mars mission worried about teamwork principles as well. Creating a suitable project organization, they noted, would be one of the "principal determinants of program cost."[9]

Small teams are less expensive to staff than large ones. So long as a project remains simple, small teams also possess an advantage in fighting failure. The mathematics of teamwork demonstrate why this is so. Small teams solve problems informally, a process that becomes more difficult as the team grows in size. The ability of a team to solve problems informally is limited by the potential number of communication channels within the team. That number increases exponentially as the membership grows in size. A 40-person team needs 4 times as many communication channels as a 20-person team. The smaller the number of communication channels, the more likely that team members can solve reliability problems without relying upon formal processes.

Complexity, both unnecessary and planned, erodes the problem-solving advantages inherent with small teams. As mission complexity grows, it quickly outstrips the ability of small teams to manage it. The complexity can be introduced by spacecraft technology or by decisions that distribute project responsibility in complicated ways.

Clearly, a limit exists to the complexity that teamwork can control. Brian Muirhead, a member of the Pathfinder team and one of the strongest advocates of "faster, better, cheaper," agrees that such limits occur. "No one yet knows what the maximum size is for an FBC project, but it's clear there is a maximum," he said. "It would be particularly foolhardy for an organization of 10,000 people to think of trying to perform an enterprise-wide Faster, Better, Cheaper project-the group is simply too large for rapid, nimble movement."[10]

Much project complexity is unnecessary and self-imposed, however, and could be reduced by more careful attention to organizational details.

CONCLUDING THOUGHTS To save time and money, advocates of "faster, better, cheaper" take risks that would seem to invite failure. They reduce redundancy; they eliminate safety features; they limit ground control. They forego many of the formal systems management procedures designed to enhance reliability. Skeptics claim that accidents will surely occur, and point to the experience of the European community as an example.

When members of the European community decided to enter the space age, they tried to build a launch vehicle called *Europa*. As a means of promoting cooperation, different nations contributed different parts. The French, Germans, Italians, and Britons manufactured components and brought them together. The results were both comical and explosive. Components from one nation would give signals to other components that caused the rockets to fly off course and explode.

The Europeans learned that they needed to adopt formal systems management in order to succeed. Individual components might be manufactured with technical precision, but until the Europeans adopted management methods capable of integrating different components, their rockets disappeared. Europeans borrowed those techniques from the U.S. space program.

Advocates of the first "faster, better, cheaper" projects departed from that well-defined mold. They used micro-technologies to produce lightweight spacecraft. They created smaller and less expensive project organizations. They avoided much of the redundancy, safety features, and instrumentation that increase spacecraft mass and complexity. Through teamwork and careful risk management, they sought to produce spacecraft that were both inexpensive and reliable.

For many years, spaceflight engineers have postulated a direct relationship between features like redundancy and spacecraft reliability. Higher spending buys redundancy which in turn increases reliability, these people say. In fact, the relationship among cost, redundancy, and reliability is not as straightforward as it seems. Additional spending does not always purchase added reliability. It buys complexity. Added complexity creates a point of diminishing returns, which eventually reduces reliability. The relationship between cost and reliability is not linear, but resembles the familiar bell-shaped curve in which the highest levels of reliability are achieved where spending rates are neither too low nor too high.

The dangers of spaceflight encourage managers to locate the appropriate balance between elements affecting low-cost innovation. Complexity, size, project organization, technology, and money affect reliability in multidimensional ways.

An analogy from the realm of terrestrial exploration may help to illustrate this point. When the first expeditions of mountaineers began to scale the highest Himalayan peaks, climbers relied upon armies of porters to establish elaborate base camps and supply caches from which a few brave souls could attempt a final ascent. This remained the conventional method for high-altitude climbing for many years. To assure success, climbing teams purchased as much equipment and as many porters as their sponsors could afford.

In 1975, Reinhold Messner and Peter Habeler pioneered a different approach. Using lightweight technology and low-cost techniques, the two climbers attempted to scale Gasherbrum I, the 26,470-foot companion to K-2, with only a single depot above a minimal base camp. At the time, few people believed that unsupported climbers—even those in superb physical condition—could scale 8 kilometer peaks. Experienced mountaineers predicted that Messner and Habler would fail to reach their destination and possibly die.

Messner and Habeler sprinted to the summit of Gasherbrum I carrying very little equipment. The techniques they perfected are now widely accepted in mountaineering circles. Confidence through repeated practice has grown. Messner pulled a sled across Antarctica to prove that the technique could be applied to other exploratory realms. Such practices are risky, but with proper preparation they need not fail.

Similar changes have allowed spacecraft managers to pursue lofty objectives with methods that are small. Lightweight technologies, small teams, simple spacecraft, and careful risk management have produced missions that work—and more than a few that failed. How far this will go is still unknown. The ultimate test would be a human expedition to Mars. It is hard to imagine such an undertaking being kept uncomplicated and small, but it also hard to imagine financing it any other way.

The leaders who won the race to the moon employed thousands of people to accomplish their goal. At the time, NASA executives boasted that the agency had assembled 300,000 people in government and industry for the purpose of conquering space. Not all of them worked on Project Apollo, but the number remained an important symbol of NASA's preference for large-scale management.

Over time, the use of thousands of people to conduct space missions may become as obsolete as the methods once used to climb Himalayan spires. As with the race to the moon, the principal barriers are not likely to be technological. They will arise from the attitudes that people bring to the challenge. The largest obstacle to low-cost innovation is the belief that it cannot be done.

NOTES

Chapter 1. The Reform

1. Joseph J. Corn, *The Winged Gospel: America's Romance with Aviation, 1900–1950* (New York: Oxford University Press, 1983).

2. Steven J. Isakowitz, Joseph P. Hopkins, and Joshua B. Hopkins, *International Reference Guide to Space Launch Systems*, 3rd ed. (Reston, Va.: American Institute of Aeronautics and Astronautics, 1999).

3. Michael Hammer and James Champy, *Reengineering the Corporation: A Manifesto for Business Revolution* (New York: Harper Business, 1993).

4. David Osborne and Ted Gaebler, *Reinventing Government: How the Entrepreneurial Spirit Is Transforming the Public Sector* (Reading, Mass.: Addison-Wesley, 1992); John M. Kamensky, "Role of the 'Reinventing Government' Movement in Federal Management Reform," *Public Administration Review* 56 (May–June 1996): 247–55.

5. Charles Perrow, *Normal Accidents: Living with High-Risk Technologies* (New York: Basic Books, 1984); Todd R. LaPorte and Paula M. Consolini, "Working in Practice But Not in Theory: Theoretical Challenges of 'High-Reliability Organizations,'" *Journal of Public Administration Research and Theory* 1, no. 1 (1991): 19–47; Martin Landau and Donald Chisholm, "The Arrogance of Optimism: Notes on Failure-Avoidance Management," *Journal of Contingencies and Crisis Management* 3 (June 1995): 67–80; Al Gore, *Creating a Government That Works Better and Costs Less: The Report of the National Performance Review* (New York: Plume, 1993).

6. Liam Sarsfield, *The Cosmos on a Shoestring: Small Spacecraft for Space and Earth Science* (Santa Monica, Calif.: RAND, 1998); David A. Bearden, "A Complexity-Based Risk Assessment of Low-Cost Planetary Missions: When Is a Mission Too Fast and Too Cheap?" paper delivered at the Fourth IAA International Conference on Low-Cost Planetary Missions, JHU/APL, Laurel, Md., May 2–5, 2000, The Aerospace Corporation, Los Angeles, Calf.

7. Quoted from Michael A. Dornheim, "Aerospace Corp. Study Shows Limits of Faster-Better-Cheaper," *Aviation Week & Space Technology*, June 12, 2000, 49.

Chapter 2. The Nature of the Challenge

1. Charles Perrow, *Normal Accidents: Living with High Risk Technologies* (New York: Basic Books, 1984).

2. NASA, Mars Program Independent Assessment Team, "Summary Report," March 14, 2000, 4; NASA, JPL Special Review Board, "Report on the Loss of the Mars Polar Lander and Deep Space 2 Missions," March 22, 2000, 114–22.

3. C. F. Larry Heimann, "Understanding the *Challenger* Disaster: Organizational Structure and the Design of Reliable Systems," *American Political Science Review* 87 (June 1993): 421–35.

4. Diane Vaughan, *The Challenger Launch Decision: Risky Technology, Culture, and Deviance at NASA* (Chicago: University of Chicago Press, 1996).

5. LaPorte and Consolini, "Working in Practice But Not in Theory."

6. Robert Pool, *Beyond Engineering: How Society Shapes Technology* (New York: Oxford University Press, 1997); Scott D. Sagan, *The Limits of Safety: Organizations, Accidents, and Nuclear Weapons* (Princeton, N.J.: Princeton University Press, 1993); C. F. Larry Heimann, *Acceptable Risks: Politics, Policy, and Risky Technologies* (Ann Arbor: University of Michigan Press, 1997); Gene I. Rochlin, *Trapped in the Net: The Unanticipated Consequences of Computerization* (Princeton, N.J.: Princeton University Press, 1997); Karlene H. Roberts, ed., *New Challenges to Understanding Organizations* (New York: Macmillan, 1993); Karl E. Weick, "Organizational Culture as a Source of High Reliability," *California Management Review* 29 (Winter 1987): 112–27; Karlene H. Roberts, "Some Characteristics of One Type of High Reliability Organization," *Organizational Science* 1, no. 2 (1990): 160–76; Karl E. Weick, Kathleen M. Sutcliffe, and David Obstfeld, "Organizing for High Reliability: Processes of Collective Mindfulness," in *Research in Organizational Behavior 1999*, vol. 21, ed. Robert I. Sutton (Greenwich, Conn.: JAI Press, 1999).

7. Stephen B. Johnson, *The Secret of Apollo: Systems Management in American and European Space Programs* (Baltimore: Johns Hopkins University Press, forthcoming).

8. Howard E. McCurdy, *Inside NASA: High Technology and Organizational Change in the U.S. Space Program* (Baltimore: Johns Hopkins University Press, 1993).

9. Martin Landau and Donald Chisholm, "The Arrogance of Optimism: Notes on Failure-Avoidance Management," *Journal of Contingencies and Crisis Management* 3 (June 1995): 67, 68.

10. Paul R. Schulman, "The Negotiated Order of Organizational Reliability," *Administration & Society* 25 (November 1993): 353–72.

11. Sarsfield, *The Cosmos on a Shoestring*, 111.

12. Mars Program Independent Assessment Team, Summary Report, 5.

13. Mars Climate Orbiter Mishap Investigation Board, "Report on Project Management in NASA," March 13, 2000, 7.

Chapter 3. Cost Control

1. Karl T. Ulrich and Steven D. Eppinger, *Product Design and Development* (New York: McGraw Hill, 1995), 5; see also Paul Mali, ed., *Management Handbook: Operating Guidelines, Techniques and Practices* (New York: John Wiley and Sons, 1981), 979–80.

2. James Webb testimony, U.S. House, Committee on Science and Astronautics, *Discussion of Soviet Man-In-Space Shot*, 87th Cong., 1st sess., April 13, 1961, 31.

3. Charles L. Schultze to Lyndon B. Johnson, Memorandum for the President, Subject: NASA Appropriations, August 11, 1967, Lyndon B. Johnson Library, WHCF EX FI 4.

4. Quoted from David Callahan and Fred I. Greenstein, "The Reluctant Racer: Eisenhower and U.S. Space Policy," in *Spaceflight and the Myth of Presidential Leadership*, ed. Roger D. Launius and Howard E. McCurdy (Urbana: University of Illinois Press, 1997), 41.

5. "An Interview with John F. Kennedy," *Bulletin of the Atomic Scientists* (November, 1960): 347; see Dodd L. Harvey and Linda C. Ciccoritti, *U.S.-Soviet Cooperation in Space* (Washington, D.C.: Center for Advanced International Studies, University of Miami, 1974), 65–79; Congressional Quarterly Service, *Congressional Quarterly Almanac 1963*, vol. 19, 172.

6. National Advisory Committee for Aeronautics to Dr. Killian's Office, August 6, 1958, NASA History Office.

7. Congressional Quarterly Service, *Congressional Quarterly Almanac 1963*, vol. 19, 172.

8. George H. Gallup, *The Gallup Poll: Public Opinion 1935–1971*, vol. 1, 1959–1971 (New York: Random House, 1972), 1720.

9. NASA, "Space Shuttle," February 1972, "Space Shuttle Economics, appendix to space shuttle fact sheet."

10. Howard E. McCurdy, "The Cost of Space Flight," *Space Policy* 10, no. 4 (1994): 284–88.

11. NASA, "Space Shuttle."

12. NASA, "Space Shuttle inflation history and effects on the agency commitment cost estimates," May 14, 1979, revised April 20, 1981; "STS non-recurring cost estimates," February 18, 1981; both in NASA History Office.

13. McCurdy, "Cost of Space Flight"; "Greatly Reduced Shuttle Cost," in Malcolm Peterson, Comptroller, NASA Headquarters, "National Aero-

nautics and Space Administration FY 2000 Budget Request Presentation," 11.

14. Roger Pielke and Radford Byerly, "The Space Shuttle Program: Performance versus Promise," in *Space Policy Alternatives*, ed. Radford Byerly (Boulder, Colo.: Westview Press, 1992).

15. Rick Gore, "When the Space Shuttle Finally Flies," *National Geographic* (March 1981): 317.

16. NASA, "Space Shuttle Program Operating Plan," fiscal year 1997. By 2000 the number of full-time equivalent contractors had been reduced to 13,350. NASA Budget Overview, Human Space Flight. Fiscal year 2000 estimates.

17. John M. Logsdon, "The Space Shuttle Decision: Technology and Political Choice," *Journal of Contemporary Business* 7, no. 3 (1978): 13–30.

18. W. R. Lucas, Program Development Memorandum to Dr. Rees, June 16, 1970, NASA History Office.

19. George M. Low, Subject: Space Vehicle Cost Improvement, May 16, 1972.

20. George M. Low, "The Cost of Doing Business in Space—A Challenge to Business and Industry," paper presented at the AIA Conference, Williamsburg, Va., May 20, 1971.

21. Quoted from Francis T. Hoban, *Where Do You Go After You've Been to the Moon? A Case Study of NASA's Pioneer Effort to Change* (Melbourne, Fla.: Krieger, 1997), 17.

22. NASA Management Instruction 129.2, September 1973.

23. Anthony Diamond, "NASA Low Cost Systems Office Benefit Analysis of Standard Equipment," March 21, 1977; see Hoban, *Where Do You Go After You've Been to the Moon?* 165–75.

24. Albert Wheelon, "Toward a New Space Policy," in *Space Policy Reconsidered*, ed. Radford Byerly (Boulder, Colo.: Westview Press, 1989), 59.

25. NASA, NASA Project Management Study, Final Oral Report, January 21, 1981, quoted from attachment entitled "Conclusions," 2, NASA History Office files.

26. Quoted from Howard E. McCurdy, *The Space Station Decision* (Baltimore: Johns Hopkins University Press, 1990), 85.

27. NASA, Office of Space Station, *The Space Station: A Description of the Configuration Established at the Systems Requirements Review* (Washington, D.C.: NASA, 1986), see p. 3. See also John J. Madison and Howard E. McCurdy, "Spending Without Results: Lessons from the Space Station Program," *Space Policy* 15 (1999): 213–21.

28. James C. Miller, Memorandum for the President, February 10, 1987, NASA History Office.

Chapter 4. The Philosophy

1. Stephanie A. Roy, "The Origin of the Smaller, Faster, Cheaper Approach in NASA's Solar System Exploration Program," *Space Policy* 14 (August 1998): 153–71.

2. "Mission to Mars May Be $300 Billion Cheaper Than First Estimates," *Aerospace Daily*, March 9, 1992, 389.

3. Roy, "Origin of the Smaller, Faster, Cheaper Approach," 163; NASA, "Report of the 90 Day Study on Human Exploration of the Moon and Mars," November 20, 1989.

4. Mark J. Albrecht, "The Council's Strategy for Space," *Roll Call*, June 25, 1990, 14.

5. Jeffrey Klein and Dan Stober, "The American Empire in Space," *San Jose Mercury News*, August 2, 1992.

6. Roy, "Origin of the Smaller, Faster, Cheaper Approach," 165; Marvin Ostrega, "Artemis Data Book, M.2.9 Clementine, Clementine Mission Timeline Summary," Artemis Society International, 1996.

7. Daniel Goldin, "Address to Employees," April 1, 1992.

8. Daniel S. Goldin, "Address to the 11th Aerospace Testing Seminar," October 10, 1988. See also, Roy, "Origin of the Smaller, Faster, Cheaper Approach," 165–66.

9. Senate Committee on Commerce, Science, and Transportation, *Nomination of Daniel S. Goldin to be Administrator of the National Aeronautics and Space Administration*, 102nd Cong., 2nd sess., March 27, 1992; Kathy Sawyer, "NASA Registers Turbulence Over New Chief's Changes," *Washington Post*, October 17, 1992.

10. Daniel Goldin, "Remarks to Jet Propulsion Lab Workers," May 28, 1992, Goldin's copy, NASA History Office.

11. Goldin, "Remarks to Jet Propulsion Lab Workers"; Goldin, "The New NASA-Faster, Better, Cheaper Without Compromising Safety," May 18, 1992.

12. Daniel S. Goldin, "From Entropy to Equilibrium: A New State of Nature for NASA," NASA Plum Brook Station, Sandusky, Ohio, October 21, 1992.

13. Goldin, "The New NASA."

14. Remarks by NASA Administrator Daniel Goldin and Discussion at the 108th Space Studies Board Meeting, November 18, 1992, Irvine, Calif., NASA History Office.

15. Goldin, "From Entropy to Equilibrium"; Goldin, "Toward the Next Millennium: A Vision for Spaceship Earth," World Space Congress, September 2, 1992.

16. Goldin, "Remarks to Jet Propulsion Lab Workers."

17. Goldin, "From Entropy to Equilibrium."

18. Goldin, "Remarks by NASA Administrator."

19. Goldin, "Remarks to Jet Propulsion Lab Workers"; Goldin, "Toward the Next Millennium"; Goldin, "Remarks by NASA Administrator."

20. Goldin, "Remarks by NASA Administrator."

21. Goldin, "Remarks to Jet Propulsion Lab Workers."

22. Goldin, "Remarks by NASA Administrator."

23. National Air and Space Museum, Department of Astronautics, Data Sheet, "Explorer-I and Jupiter-C"; NASA, John F. Kennedy Space Center, "10th Anniversary Launch of Explorer I," January 31, 1968.

24. James R. Wertz and Wiley J. Larson, eds., *Reducing Space Mission Cost* (Torrance, Calif.: Microcosm Press, 1996), 6–8.

25. See Craig Covault, "Major Space Effort Mobilized to Blunt Environmental Threat," *Aviation Week & Space Technology*, March 13, 1989, 36–44; James R. Asker, "Earth Mission Faces Growing Pains," *Aviation Week & Space Technology*, February 21, 1994, 36.

26. Roy, "Origin of the Smaller, Faster, Cheaper Approach," 157.

27. Norman R. Augustine, *Augustine's Laws* (New York: Penguin Books, 1986), 143.

28. National Research Council, *Reducing the Costs of Space Science Research Missions* (Washington, D.C.: National Academy Press, 1997), 30–33.

29. Quoted from NASA, "NASA's Discovery Program: Solar System Exploration for the Next Millennium," n.d.

30. NASA, NASA News, "NASA Selects 11 Discovery Mission Concepts for Study," release 93–027, February 11, 1993; NASA, NASA News, "NASA Names Science Team for Asteroid Rendezvous Mission," release 94–159, September 21, 1994.

31. NASA, "NASA's Discovery Program."

32. Helen Worth, "NEAR Team Recovers Mission After Faulty Engine Burn," February 4, 1999, *near.jhuapl.edu.*

33. NASA, "1998 Mars Missions," press kit, December 1998, 12.

34. NASA Goddard Space Flight Center, NASA Facts, "NASA's Small Explorer Program: Faster, Better, Cheaper," January 1998, 1; NASA Facts on Line, NASA Headquarters, "Space Technology and Technology Development," February 4, 1999.

35. NASA, News Release, "NASA Terminates Clark Earth Science Mission," release 98–35, February 25, 1998; Andrew Lawler, "Faster, Cheaper Strategy on Trial," *Science,* November 14, 1997, 1216–17.

36. See Warren Ferster, "Cost Overruns May Kill NASA's Clark Satellite," *Space News,* April 21–27, 1997.

37. Goldin, "Remarks by NASA Administrator."

38. Robert L. Abramson, David A. Bearden, and David L. Glackin, "Small Satellites: Cost Methodologies and Remote Sensing Issues," in

Advanced and Next-Generation Satellites, ed. Hiroyuki Fujisada and Martin N. Sweeting (Bellingham, Wash.: SPIE—The International Society for Optical Engineering, 1995), 558.

Chapter 5. Mars *Pathfinder*

1. Between the eras of the *Viking* and *Pathfinder* probes, the price of aerospace purchases increased by a factor of 3.7 due to inflation. NASA, Office of Chief Financial Officer, "NASA New Start Inflation Index," April 29, 1998.

2. "NASA Seeks 1994 Start for Mars Lander Mission," *Aviation Week & Space Technology,* June 29, 1992, 49.

3. Specifically, the Discovery program (which Mars *Pathfinder* was part of) required project managers to restrict spending to $150 million in fiscal year 1992 dollars for "project development costs, which includes final design, fabrication, assembly, and test through launch plus 30 days." The cap did not include "the cost of advanced studies, project definition, launch vehicles, and mission operations/data analysis." Pathfinder development expenses totaled $170 million in actual spending, or $155 million in 1992 dollars. NASA, "NASA's Discovery Program: Solar System Exploration for the Next Millennium," n.d., 1.

4. A *Delta* 2-7925 rocket cost approximately $50 million to deliver 11,000 pounds to low Earth orbit. Frank Sietzan, *World Guide to Commercial Launch Vehicles* (Arlington, Va.: Pasha Publications, 1991); Isakowitz, Hopkins, and Hopkins, *International Reference Guide to Space Launch Systems.* As of 1992, the space shuttle delivered as much as 55,000 pounds to low Earth orbit at a cost of $393 million per flight. McCurdy, "Cost of Space Flight."

5. NASA, "1996 Mars Missions," press kit, November 1996.

6. "Viking Mission to Mars," from House Subcommittee on Space Science and Applications, Science and Astronautics Committee, *Viking Project,* 93rd Cong., 2nd sess., 1974, 35–36.

7. David Pieri, "Mars Pathfinder: Looking Back on The Little Mission that Could," *Launchspace Magazine* (December 1997–January 1998): 22–25.

8. House Subcommittee on Space Science and Applications, *Viking Project,* 196.

9. Edward C. and Linda N. Ezell, *On Mars: Exploration of the Red Planet, 1958–1978,* NASA SP-4212 (Washington, D.C.: U.S. Government Printing Office, 1984), 251.

10. Priestley Toulmin, "You Get What You Pay For," *Washington Post,* July 22, 1997, A.14.

11. Ezell and Ezell, *On Mars,* 251.

12. Ibid., 214; Thomas A. Mutch, *The Martian Landscape* NASA SP-425 (Washington, D.C.: U.S. Government Printing Office, 1978), 17.

13. Ezell and Ezell, *On Mars,* 214, 231, 251.

14. Donna Shirley, *Managing Martians* (New York: Broadway Books, 1998), 183.

15. Ezell and Ezell, *On Mars,* 269.

16. Ibid., 451; Martin Marietta Corporation, *The Viking Mission to Mars* (Denver: Martin Marietta Corporation, Public Relations Department, 1975), sec. III, 87.

17. Ezell and Ezell, *On Mars,* 447.

18. NASA, "Budget Estimates: Fiscal Year 1996," vol. II, sec. 3, 1–10.

Chapter 6. Organization

1. Quoted from Stephen B. Johnson, "Craft or System? The Development of Systems Engineering at JPL," *Quest* 6 (Summer 1998): 24. For more information on the *Ranger* probe, see House Committee on Science and Astronautics, Subcommittee on NASA Oversight, *Investigation of Project Ranger,* 88th Cong., 2nd sess., 1964, and R. Cargill Hall, *Lunar Impact: A History of Project Ranger,* SP-4210 (Washington, D.C.: U.S. Government Printing Office, 1977).

2. Stephen B. Johnson, "Systems Management in the American and European Space Programs," manuscript to be published as *The Secret of Apollo* by Johns Hopkins University Press, 91.

3. Stephen B. Johnson, "The Organizational Roots of American Economic Competitiveness in High Technology," *Space Times* (March–April 1999): 10. See also John C. Lonnquest, *The Face of Atlas: General Bernard Schriever and the Development of the Atlas Intercontinental Ballistic Missile, 1953–1960,* dissertation submitted in partial fulfillment of the requirements for the degree of Doctor of Philosophy in the Department of History in the Graduate School of Duke University, 1996.

4. Charles E. Cockrell, "Lessons Learned from Better, Faster, Cheaper Concepts As Applied to Selected NASA Programs," Enterprise Safety and Mission Assurance Division, Office of Safety and Mission Assurance, NASA Headquarters, Washington, D.C., November 5, 1998, 4.

5. Donna Shirley, *Managing Martians* (New York: Broadway Books, 1998), 180, 274.

6. Mars Pathfinder Roundtable, November 5, 1998.

7. See Andrew Lawler, "Faster, Cheaper Strategy on Trial," *Science* 278 (November 14, 1997) and Warren Leary, "NASA Learns That Faster and Cheaper Isn't Always So," *New York Times,* September 15, 1998.

8. Mars Pathfinder Roundtable.

9. Price Pritchett and Brian Muirhead, *The Mars Pathfinder Approach to "Faster-Better-Cheaper"* (Dallas: Pritchett & Associates, 1998), 70.

10. Mars Pathfinder Roundtable.

11. Shirley, *Managing Martians*, 197-99.

12. Mars Pathfinder Roundtable.

13. See Gareth Morgan, *Images of Organization,* 2nd ed. (Thousand Oaks, Calif.: Sage Publications, 1997), 112-13.

14. See Perrow, *Normal Accidents,* 332-34.

15. See Thomas J. Lewin and V. K. Narayanan, "Keeping the Dream Alive: Managing the Space Station Program, 1982-1986," NASA Contractor Report 4272, July 1990, 52.

16. See Howard E. McCurdy, *The Space Station Decision: Incremental Politics and Technological Choice* (Baltimore: Johns Hopkins University Press, 1990), 204-12.

17. McCurdy, *Inside NASA,* 130.

18. NASA, Mars Program Independent Assessment Team, "Summary Report," March 14, 2000, 12.

Chapter 7. Technology

1. Carl A. Kukkonon, JPL Center for Space Microelectronics, *Microelectronics for a New Millennium in Space Missions,* videotape produced at the Langley Research Center, February 23, 1996.

2. NASA Facts, "Center for Space Microelectronics Technology," Jet Propulsion Laboratory, Pasadena, Calif., 3.

3. See Larry Armstrong and Larry Holyoke, "NASA's Tiny Camera Has a Wide-Angle Future," *Business Week,* March 6, 1995.

4. Quoted from William E. Burrows, "The New Millennium," *Air & Space* (August-September 1996): 50.

5. NASA, "1998 Mars Missions: Press Kit," December 1998, 58.

6. NASA, Mars Program Independent Assessment Team, 5.

7. NASA, "Deep Space 1 Launch: Press Kit," October 1998, 27.

8. Christine Anderson, chair, "Lewis Spacecraft Mission Failure Investigation Report: Final Report," February 12, 1998.

Chapter 8. Risk and Reliability

1. Goldin, "Remarks to Jet Propulsion Lab Workers," "From Entropy to Equilibrium."

2. Senior Management Teleconference, NASA Headquarters, February 7, 2000.

3. Quoted from McCurdy, *Inside NASA,* 71; NASA, "Mars Pathfinder Frequently Asked Questions: Entry, Descent and Landing," April 10, 1996.

4. Aaron Wildavsky, *Searching for Safety* (New Brunswick: Transaction Books, 1988), 205.

5. See Michael A. Dornheim, "Airbag Passes Test on Mars," *Aviation Week & Space Technology*, July 14, 1997, 36; also Shirley, *Managing Martians*, 199–201.

6. NASA, "Mars Pathfinder Frequently Asked Questions: Entry, Descent and Landing."

7. William Boyer, "Despite Success, Airbag Future Unclear," *Space News*, July 14–20, 1997.

8. Correspondence with Sam W. Thurman, Jet Propulsion Laboratory, March 21, 1999.

9. G. Scott Hubbard, "Lunar Prospector: Developing a Very Low Cost Planetary Mission," unpublished paper 175–2, n.d.; Brian K. Muirhead and William L. Simon, *High Velocity Leadership* (New York: HarperBusiness, 1999), 25.

10. See Time-Life Books, *Space: Understanding Computers* (Alexandria, Va.: Time-Life Books, 1987).

11. Mars Pathfinder Roundtable.

12. Pritchett and Muirhead, *The Mars Pathfinder Approach to "Faster-Better-Cheaper,"* 58.

13. Mars Pathfinder Roundtable.

14. Ibid.

15. Pritchett and Muirhead, *The Mars Pathfinder Approach to "Faster-Better-Cheaper,"* 65.

16. Quoted from Scott D. Sagan, *The Limits of Safety: Organizations, Accidents, and Nuclear Weapons* (Princeton, N.J.: Princeton University Press, 1993), 47.

17. See Vaughan, *The Challenger Launch Decision;* Charles Perrow, "The Habit of Courting Disaster," *Nation* 243 (October 11, 1986): 347–56.

18. Sarsfield, *The Cosmos on a Shoestring;* Bearden, "A Complexity-Based Risk Assessment of Low-Cost Planetary Missions."

19. Sarsfield, *The Cosmos on a Shoestring,* 107.

Chapter 9. Future Implications
1. "The Rationale for Infrared Astronomy: Clever Choices," SIRTF web site, *sirtf.caltech.edu.*

2. Daniel S. Goldin, July 2, 1996.

3. Quoted from Lee Siegel, "Spaceplane Is Delayed by Fuel Tank Flaw," *Salt Lake Tribune*, January 22, 1999.

4. Quoted from Jeff Leeds, "Development of Reusable Rocket Delayed," *Los Angeles Times*, January 19, 1999.

5. Quoted from Seth Borenstein, "Taking that Trip to Mars," *San Jose Mercury News*, December 15, 1998.

6. Quoted from Seth Borenstein, "Testing Flights to Mars," *Portland Oregonian*, December 27, 1998.

7. Theresa Foley, "NASA Team Modifies Mars Direct Mission," *Space News*, January 30-February 5, 1995; David Chandler, "Mars on $5m a day: Well, not exactly, but a lot cheaper (and faster) than we thought," *Boston Globe*, August 5, 1996.

8. Quoted from Frank Sweeney and Glennda Chui, "Space Flight on the Cheap," *San Jose Mercury News*, March 2, 1999.

9. NASA, "Human Exploration of Mars: The Reference Mission of the NASA Mars Exploration Study Team," ed. Stephen J. Hoffman and David L. Kaplan, July 1977.

10. Mars Pathfinder Roundtable; Muirhead, *High Velocity Leadership*, 225.

INDEX

Abrahamson, James, 47
accountability, 91–93, 100
Albrecht, Mark, 45–46
Allen, Paul, 91
Apollo applications program, 34
Apollo project, 4, 9, 18, 42, 156; analog for Mars mission, 150–51; *Apollo 13*, 18; *Apollo 204* fire, 13, 38; cost, 4, 32, 148–49; management methods, 87, 103
Applied Physics Laboratory, 45, 93, 95
Atlas missile program, 84; *Atlas-Agena* launch vehicle, 81–82
Augustine, Norman, 54–55

Bearden, David, 9, 26–27; "Bearden rule," 11
Beggs, James, 42, 103
"better." *See* reliability
Brilliant Pebbles concept, 46, 48
Bush, George, 44–45, 148

California Institute of Technology, 80
capability of spacecraft, 51, 67–70, 144
Carnegie Mellon University, 117
Cassini mission, 3, 19, 97, 99, 104, 106–7, 109, 111, 119, 141; cost, 106, 120–21, 140; weight, 121, 139–40
Centaur rocket, 64
Center for Space Microelectronics, 108
Challenger space shuttle, 13, 19
Chandra X-Ray Observatory, 140, 142–44
cheap access to space. *See* launch costs
"cheaper." *See* cost of space flight; cost reduction
China Lake Naval Weapons Center, 132

Clark Earth Observing Satellite, 5, 7, 57–59, 95
Class 3 (or Class C) satellites, 59–60, 125, 134, 142
Clementine project, 46–47, 55
Cockrell, Charles, 79, 94–95
Collins, Michael, 36
co-location of project team, 24, 29, 93–94
common components, 4, 38–40, 51, 100–101
complexity, 101–2, 153; and coordination, 102–5; defined, 16–17, 134–35; relation to cost and/or schedule, 26–27, 139–41, 155; requisite variety, 101; on spacecraft, 26
Compton Gamma Ray Observatory, 143
Concorde supersonic transport, 13
concurrency, 85
configuration management, 85–86
contracting out, 28–29, 95, 104
cost of space flight, 3–4, 19, 54, 104–5, 120, 128, 140–43; Apollo project, 4, 32, 148–49; aviation analogy, 2–3; Cassini mission, 106, 120–21, 140; cost reduction in society, 1, 138; culture of cornucopia, 31–34, 38, 42; Discovery program, 56; International Space Station, 40–43, 149–50; launch costs, 3, 63–64, 144–45, 149–50; Mars expedition (human), 148–51; Mars *Global Surveyor*, 69–70, 104, 120, 140; Mars *Observer*, 18, 120, 140; Mars *Pathfinder*, 62–64, 67–68, 70, 76–78, 104, 120–22, 127–28, 140–41; Mars *Polar Lander*, 19, 104–5, 120, 128; relation to complexity, 26–27, 139–41,

Printed in the United States
1524700001BA/70-78